# TRUST IN MILITARY TEAMS

# Human Factors in Defence

*Series Editors*:

Dr Don Harris, Managing Director of HFI Solutions Ltd, UK
Professor Neville Stanton, Chair in Human Factors at the University of Southampton, UK
Dr Eduardo Salas, University of Central Florida, USA

Human factors is key to enabling today's armed forces to implement their vision to 'produce battle-winning people and equipment that are fit for the challenge of today, ready for the tasks of tomorrow and capable of building for the future' (source: UK MoD). Modern armed forces fulfil a wider variety of roles than ever before. In addition to defending sovereign territory and prosecuting armed conflicts, military personnel are engaged in homeland defence and in undertaking peacekeeping operations and delivering humanitarian aid right across the world. This requires top class personnel, trained to the highest standards in the use of first class equipment. The military has long recognised that good human factors is essential if these aims are to be achieved.

The defence sector is far and away the largest employer of human factors personnel across the globe and is the largest funder of basic and applied research. Much of this research is applicable to a wide audience, not just the military; this series aims to give readers access to some of this high quality work.

Ashgate's *Human Factors in Defence* series comprises of specially commissioned books from internationally recognised experts in the field. They provide in-depth, authoritative accounts of key human factors issues being addressed by the defence industry across the world.

# Trust in Military Teams

EDITED BY

NEVILLE A. STANTON
*University of Southampton, UK*

CRC Press
Taylor & Francis Group
Boca Raton London New York

CRC Press is an imprint of the
Taylor & Francis Group, an **informa** business

CRC Press
Taylor & Francis Group
6000 Broken Sound Parkway NW, Suite 300
Boca Raton, FL 33487-2742

First issued in paperback 2017



Version Date: 20160226

ISBN 13: 978-1-138-07594-8 (pbk)
ISBN 13: 978-1-4094-0448-4 (hbk)



**Visit the Taylor & Francis Web site at**
**http://www.taylorandfrancis.com**

**and the CRC Press Web site at**
**http://www.crcpress.com**

# Contents

# List of Figures

# List of Tables

# Preface

Trust is has always been an important topic for military teams, as effective war-fighting means reliance upon the cooperation and collaboration of others. Trust seems especially important in contemporary operations for joint, multinational, coalition forces who are working with new technology and undertaking stabilisation and reconstruction activities. This is the focus of the book, reporting on investigations into trust between people, between and within teams, between and within organisations, and trust in technology. The book was conceived at a meeting of The Technical Cooperation Programme (TTCP: see http://www.dtic.mil/ttcp/) Human Resources and Performance Group Technical Panel 11 (HUM TP11) on Human Aspects of Command (chaired by Dr Joan Johnson (USA) with panel members Dr Jay Goodwin (USA), Dr Dan McDonald (USA), Dr Beejal Mistry (UK), Dr Matthew Duncan (Canada), L.Col Jeff Stouffer (Canada), and Dr Glen Smith (Australia). It was agreed that all of the TTCP nations would contribute chapters indicative of some of the work that they were conducting on trust in military teams. By way of a brief historical background, TTCP formally came to life on 25 October 1957, when the then President of the United States and Prime Minister of Great Britain made the following statement:

> The arrangements which the nations of the free world have made for collective defence and mutual help are based on the recognition that the concept of national self-sufficiency is now out of date. The countries of the free world are interdependent and only in genuine partnership, by combining their resources and sharing tasks in many fields, can progress and safety be found. For our part we have agreed that our two countries will henceforth act in accordance with this principle.

The USA and UK were joined by Canada to form the Tripartite Technical Cooperation Programme. Australia joined in 1965, and New Zealand joined in 1969. These five nations form the current membership, and the organization now called The Technical Cooperation Programme (TTCP). The twin ideas of collective defence and mutual help are just as pertinent today as they were over 50 years ago, as we are observing in stabilisation, peacekeeping and disaster relief operations all over the world. It is a fitting tribute to these joint, multinational, coalition operations that HUM TP11 has collectively produced material for this book, so that we may benefit from the research of each nation. The book begins with an introduction to the domain as well as providing an overview and summary of the contributions. The book has been divided into three main sections: Experimental Studies into Mood, Personality and Training; Multinational Challenges for Trust;

and Trust in Technology. It is expected that the reader will select the sections relevant to their work, dipping into chapters as appropriate.

Professor Neville A. Stanton, PhD.
University of Southampton
October 2010

# Biographies of Authors

**C. Shawn Burke** is a Research Scientist at the Institute for Simulation and Training of the University of Central Florida. Her expertise includes teams and their leadership, team adaptability, team training, measurement, evaluation, and team effectiveness. Dr Burke has published over 60 journal articles and book chapters related to these topics and has presented/had work accepted at over 100 peer-reviewed conferences. She is currently investigating team adaptability as well as issues related to multicultural team performance and multiteam systems. All of this work is conducted with an interest in team leadership and the training of teams operating in complex environments. Dr Burke earned her doctorate in industrial/organizational psychology from George Mason University. Dr Burke serves as an ad-hoc reviewer for *Human Factors*, *Leadership Quarterly*, *Journal of Applied Psychology*, *Human Resource Management*, *Journal of Occupational and Organizational Psychology*, and *Quality and Safety in Healthcare*. She has co-edited a book on adaptability and one on advances in team effectiveness research.

**Anna T. Cianciolo** is the President and Senior Behavioral Scientist at Command Performance Research, Inc. Since earning her PhD in Engineering Psychology (Georgia Tech, 2001), she has investigated the technical, social, and organizational requirements for effectively implementing collaborative information systems for Army education and operations. She has explored such topics as assessing interpersonal trust in networked teams, using 3D-animated videos to stimulate discussion in online professional forums, designing and implementing distributed, collaborative negotiation training, and developing metrics for evaluating blended learning. Her postgraduate studies included a two-year postdoctoral research position at the Yale University Center for the Study of the Psychology of Abilities, Competency, and Expertise (PACE Center). While at the PACE Center, she studied performance development with a focus on experience-based, or tacit, knowledge and military leadership. Dr Cianciolo is also an Adjunct Assistant Professor in the Department of Educational Psychology, School of Education at the University of Illinois, Urbana-Champaign. She has published her work in books, edited volumes, and journals and has been an invited speaker at numerous conferences. She is on the editorial board of the *Journal of Experimental Psychology: Applied.*

**Arwen H. DeCostanza** is a Research Psychologist at the US Army Research Institute for the Behavioral and Social Sciences (ARI) in the Aberdeen Proving Ground, MD field office. Dr DeCostanza received her PhD in Industrial and Organizational Psychology from The George Washington University in June 2009. While completing her doctorate, she worked in ARI's Personnel Assessment

Research Unit as a Consortium Research Fellow, conducting research focused on the development and validation of non-cognitive selection measures. After completing her degree, Dr DeCostanza moved to the Aberdeen field office and began research focused on enhancing performance in complex teams, including joint, interagency, multinational, and distributed teamwork. Her research explores topics related to team decision making, trust in distributed teams, team consensus building, information sharing and collaboration, culture, and cohesion in relation to team performance and adaptability. She has produced numerous conference presentations and publications covering these topics. Dr DeCostanza is a member of the Society for Industrial and Organizational Psychology, American Psychological Association, and the Interdisciplinary Network for Group Research.

**Karen M. Evans** is a cognitive scientist with broad experience in both basic and applied aspects of human learning. As a Junior Behavioral Researcher at Command Performance Research, Inc., she examined the exchange of knowledge in distributed groups, emphasizing the role of trust in facilitating this process. This work was largely conducted in support of the company's Army Research Institute-funded project on the development of interpersonal trust in network-enabled military teams. Dr Evans is currently exploring applications of eye tracking methods as a postdoctoral researcher at the Rochester Institute of Technology. Much of her research examines perceptual learning in higher education, and she is working with faculty in multiple disciplines to incorporate these findings into instructional practices. She is also working with corporate sponsors to apply studies of visual attention to the design and marketing of consumer products. Dr Evans has published numerous experiments on human memory; many of these studies investigate how stimulus-specific details are encoded and retrieved, and also explore the sensitivity of these processes to age-related changes and strategy selection. She holds BAs in Psychology and in English from Kenyon College, and an MS and PhD in Cognitive Psychology from the University of Illinois at Urbana-Champaign.

**Angela R. Febbraro** received her PhD in Applied Social Psychology and her MA in Industrial/Organizational Psychology at the University of Guelph, Canada. Dr Febbraro is currently a defence scientist at Defence Research & Development Canada (DRDC) – Toronto, where she is Group Leader of the Organizational Behavior Group within the Collaborative Performance and Learning Section. Over the course of her career, Dr Febbraro has published over 40 peer-reviewed publications and delivered over 70 oral presentations at national and international conferences and symposia. Her publications and presentations in military psychology span the topics of cultural diversity, gender, and reintegration issues for military personnel, and her current research focuses on social and organizational aspects of effective civil-military interaction in international military engagements, including the role of organizational culture, stereotypes, and trust. A former faculty member in community psychology at Wilfrid Laurier University, Dr Febbraro currently holds an adjunct faculty position in psychology at York University, and has taught several courses in the psychology of women, the psychology of gender, industrial/organizational psychology, and social psychology. Dr Febbraro has served on the editorial board of a number of journals, including the *Canadian Journal of Police and Security Services*, *Theory and Psychology*, and the *History and Philosophy of Psychology Bulletin*. Dr Febbraro has been the recipient of a number of scholarships and awards for her work, from the Canadian Psychological Association, the University of Guelph, and the University of Toronto.

**Stephen M. Fiore** is Associate Professor with the University of Central Florida's Cognitive Sciences Programme in the Department of Philosophy and Director of the Cognitive Sciences Laboratory at UCF's Institute for Simulation and Training. He earned his PhD degree in Cognitive Psychology from the University of Pittsburgh, Learning Research and Development Center (2000). Dr Fiore maintains a multidisciplinary research interest that incorporates aspects of the cognitive, computational, and organizational sciences in the investigation of learning and performance in individuals and teams. He is co-editor of recent volumes on macrocognition in teams (2008), distributed learning (2007), and team cognition (2004), and he has co-authored over 100 scholarly publications in the area of learning, memory, and problem solving at the individual and the group level. As Principal Investigator and Co-Principal Investigator he has helped to secure and manage approximately $15 million in research funding from organizations such as the National Science Foundation, the Office of Naval Research, the Air Force Office of Scientific Research, and the Department of Homeland Security.

**Elisabeth W. Fitzhugh** is a cognitive scientist with over 20 years' experience, currently employed by SRA International, Inc. Ms. Fitzhugh's primary areas of expertise are requirements development for both system design and training and experimental design for human performance tests and usability evaluation. Special interests include socio-cultural issues in human error, individual difference in decision processes, and trust issues in joint socio-technical systems. Recent work includes participation in an exploratory trust workshop conducted by Air Force Research Laboratory's 711th HPW and a trust-oriented usability assessment of the Information Directorate's National Operational Environment Model (NOEM). Other projects include assessing socio-cultural effects on sensemaking and reliability in electrical power grid management; initiating a National Level Strategy Planning Workshop which culminated in a gap analysis of support to Joint Planning; and assessing information exchange issues for the Air Operations Center Strategy Division. Ms Fitzhugh has written articles on 3D visualizations in military applications and on the issues specific to trust in joint socio-technical systems. Ms Fitzhugh received her BS from Wright State University. She is member of the Human Factors and Ergonomics Society and of the Animal Behavior Society.

**Gregory J. Funke** is a Research Psychologist in the US Air Force Research Laboratory's 711th Human Performance Wing. He has published over 25 conference papers, book chapters, and journal articles in the areas of human factors and neuroergonomics. Dr Funke received his PhD in Human Factors Psychology from the University of Cincinnati, and completed a postdoctoral research fellowship with the Consortium Research Fellows Programme at AFRL.

**Sena Garven** is currently a Research Psychologist at the US Army Research Institute, Ft. Leavenworth Research Unit. Her primary research interests are self-awareness and tacit knowledge for military leaders. She is on the editorial board of *Military Psychologist*.

**Ritu Gill** is a Defence Scientist at Defence Research & Development Canada – Toronto (DRDC Toronto). As part of the Collaborative Performance and Learning Section at DRDC Toronto, she is conducting research that examines trust violations and trust restoration between military organizations and indigenous populations as a means of enhancing mission success in international military engagements. She has published several peer-reviewed publications in the area of trust restoration, sexual harassment, and gender discrimination. Dr Gill has served as a peer reviewer for the American Psychological Association *Journal of Personality and Social Psychology Bulletin.* She has also taught several introductory psychology courses at Carleton University. Dr Gill received her PhD in Social Psychology from Carleton University in 2007.

**Robert R. Hoffman** is a Senior Research Scientist at the Institute for Human and Machine Cognition in the USA. He is a Fellow of the Association for Psychological Science and a Fulbright Scholar. His PhD is in Experimental Psychology from the University of Cincinnati, where he received McMicken Scholar, Psi Chi, and Delta Tau Kappa Honors. Following a postdoctoral associateship at the Center for Research on Human Learning at the University of Minnesota, Dr Hoffman joined the faculty of the Institute for Advanced Psychological Studies at Adelphi University. He began his career as a psycholinguist, and founded the journal *Metaphor and Symbol*. His subsequent research leveraged the psycholinguistics background in the study of methods for eliciting the knowledge of domain experts. Dr Hoffman has been recognized internationally in disciplines including psychology, remote sensing, weather forecasting, and artificial intelligence, for his research on human factors in remote sensing, his work in the psychology of expertise and the methodology of cognitive task analysis, and for his work on HCC issues intelligent systems technology and the design of macrocognitive work systems. Dr Hoffman is a co-editor for the Department on Human-Centered Computing in IEEE: Intelligent Systems. He is editor for the book series 'Expertise: Research and Applications'. He is a co-founder and track editor for the *Journal of Cognitive Engineering and Decision Making*. His major current projects involve evaluating the effectiveness of knowledge management, and performance measurement for macrocognitive work systems.

**Justin G. Hollands** is a Defence Scientist and Senior Advisor to the Human Systems Integration Section at Defence Research & Development Canada – Toronto (DRDC Toronto). He served as Head of the Human Systems Integration Section at DRDC Toronto from 2006–10, and was Acting Chief Scientist for DRDC Toronto in 2008. Dr Hollands has been a Defence Scientist at DRDC Toronto since 1999. Prior to that he was Assistant Professor in the Department of Psychology at the University of Idaho from 1994 until 1999, where he headed the Human Factors Graduate Programme. Dr Hollands is Adjunct Associate Professor in the Department of Mechanical and Industrial Engineering at the University of Toronto, where he is also an associate member of the School of Graduate Studies. He is an associate editor of *Ergonomics in Design*. Dr Hollands has authored or co-authored over 50 scientific articles and is co-author of *Engineering Psychology and Human Performance* (3rd edn), with Christopher Wickens. Dr Hollands has an extensive background in the human factors of display design, spatial cognition, perception and psychophysical scaling, information visualization, and cognitive ergonomics. He received a PhD in Psychology from the University of Toronto in 1993.

**Sandra Hughes** is a Senior Research Psychologist in the Training and Human Performance Research & Development Branch at the Naval Air Warfare Center, Training Systems Division, in Orlando, Florida. Ms. Hughes received an MS in Industrial and Organizational Psychology and a BS in Psychology, with a minor in Business from the University of Central Florida in Orlando. She has conducted and/or managed research and development projects in the areas of: social psychology, team performance, stress and decision making, and distance learning design and implementation. She has published papers in peer-reviewed journals, proceedings, and books on military training methods and technologies, culture, and stress. She is currently working primarily in the area of mentoring (implementation and support) and culture: design and development of cultural training, measurement of cross-cultural competence, and evaluation of cultural training. Ms. Hughes' research interests include the areas of trust and culture: the effects of cultural distance on trust, how culture affects the demonstration and evaluation of trustworthiness cues, how cultural biases affect the trust building process, and how trust can be repaired between culturally distant parties.

**Dan Jenkins** graduated in 2004, from Brunel University, with an M.Eng (Hons) in Mechanical Engineering and Design, receiving the 'University Prize' for the highest academic achievement in the school. As a sponsored student, Dan finished university with over two years' experience as a Design Engineer in the automotive industry. Upon graduation, Dan went to work in Japan for a major car manufacturer, facilitating the necessary design changes to launch a new model in

Europe. In 2005, Dan returned to Brunel University taking up the full-time role of Research Fellow in the Ergonomics Research Group, working primarily on the HFI-DTC project. Dan studied part-time on his PhD in Human Factors and Interaction Design – graduating in 2008, receiving the 'Hamilton Prize' for the Best Viva in the School of Engineering and Design. Both academically and within industry Dan has always had a strong focus on human factors, system optimization and design for inclusion. Dan has authored and co-authored numerous journal paper, conference articles, book chapters and books. Dan and his colleagues on the HFI DTC project were awarded the Ergonomics Society's President's Medal in 2008.

**Joan Johnston** has 25 years of US government civilian service and has been employed as a Research Psychologist with the US Navy for 20 years. Her primary areas of expertise include training and decision support systems for tactical decision making under stress, team performance and team training technologies, embedded and distributed simulation-based training technologies, leadership and operational readiness in joint and multinational exercises, and cross-cultural competence. She has published over 60 papers in peer-reviewed journals, proceedings, and books on military training methods and technologies. For her contribution to the Office of Naval Research Programme on Tactical Decision Making Under Stress, she received the Dr Arthur E. Bisson Prize for Naval Technology Achievement (2000), and the Society for Industrial and Organizational Psychology M. Scott Myers Award for Applied Research in the Workplace (2001). In recognition for outstanding research and technical achievements to naval aviation, Dr Johnston was inducted into the Naval Air Warfare Systems Command Research and Engineering Fellows Programme in 2000. In recognition for her sustained excellence in scientific achievements she was promoted to Fellow status in 2008. Dr Johnston holds a BS (Hons) degree from Rutgers University, and an MA and PhD in Industrial and Organizational Psychology from the University of South Florida.

**Benjamin A. Knott** has a general interest in behavioural science as it relates to the design of human–machine interfaces. His current emphasis is on studying team-based metrics of workload and performance, and the impact of technology-mediated communication on tactical command and control teams. He currently holds a position as Research Psychologist and Section Chief at the US Air Force Research Lab's 711th Human Performance Wing. Dr Knott's past R&D activities include the design of natural language user interfaces for automated systems, the examination of virtual reality as an advanced interface for military training and mission rehearsal, and the study of virtual environments for assessment and rehabilitation of neuropsychological impairments. Dr Knott has published over 50 conference papers, book chapters and journal articles in the areas of human factors, usability engineering, and neuroergonomics. Additionally, he has over 40 US and international patents on inventions concerning user interfaces, automated systems, speech technologies and communication systems. Dr Knott has a PhD in Applied Experimental Psychology from The Catholic University of America. He also held a graduate research fellowship for the US Army Research Institute and a postdoctoral research fellowship at The National Rehabilitation Hospital's Assistive Technology Research Center.

**Joseph B. Lyons** is the Team Lead for the Organizational Effectiveness research area within the 711th Human Performance Wing – Sensemaking and Organizational Effectiveness Branch (711th HPW/RHXS) at Wright-Patterson Air Force Base in Dayton, Ohio. Dr Lyons received his PhD in Industrial/Organizational Psychology from Wright State University in Dayton, Ohio, in 2005. Some of Dr Lyons' research interests include trust, leadership, stress and emotions, organizational assessment and organizational change. Dr Lyons is currently researching the social and other contextual predictors of trust in decision making and team settings. Dr Lyons has published numerous conference papers, book chapters, and journal articles in the areas of trust, organizational change, and leadership. Dr Lyons is currently serving as the Trust Research Portfolio Lead for the Human Effectiveness Directorate.

**Anthony Masys** is a Defence Scientist with Defence Research & Development Canada (DRDC) Centre for Security Science and is currently assigned to the Royal Canadian Mounted Police (RCMP) as Scientific Advisor. In support of RCMP, he has been involved providing scientific advice and analysis during the Vancouver 2010 Olympics and G8/G20 Summits in 2010. A former military aviator, Dr Masys received his BSc in Physics and MSc in Underwater Acoustics and Oceanography from the Royal Military College of Canada (RMC) and his PhD from the University of Leicester, UK (focusing on human error and complex socio-technical systems). He is currently supporting the distance MSc Programme in Risk, Crisis and Disaster Management at the University of Leicester as tutor and lecturer.

**Cecily E. E. McCoy** is a Research Psychologist in the Training and Human Performance Research & Development Branch at the Naval Air Warfare Center Training Systems Division in Orlando, Florida. She is currently working on programs related to the cross-cultural relations and trust for military teams. Her Belizean-American bicultural heritage provides a unique perspective on cultural issues. Ms. McCoy is especially interested in investigating the relation between multicultural identity and cultural adaptability. She holds a BA in Psychology (Magna Cum Laude) and an MS in Sport Management from Florida International University. She continued her education at the University of Central Florida (UCF) where she earned an MS degree in Industrial and Organizational Psychology. During her graduate career, Ms. McCoy has taught courses on cross-cultural psychology and the psychology of racial prejudice at UCF. She was awarded the McKnight Doctoral Fellowship and is currently pursuing her PhD in Industrial and Organizational Psychology from UCF, with her dissertation topic being a multinational study examining the cultural influence on information sharing and team performance.

**Janet E. Miller** is a Senior Electronics Engineer and Program Manager in the US Air Force (USAF) Research Laboratory's Sensors Directorate at Wright-Patterson Air Force Base, Ohio. Dr Miller received a PhD in Industrial and Systems Engineering from The Ohio State University in 2002 with emphasis on cognitive aspects of human factors. Since receiving her degree, she has focused her efforts on addressing various cognitive challenges of work, including teamwork, in US Air Operations Centers (AOC) in particular but in other USAF settings as well. To support that work, she has visited AOCs worldwide, documenting the challenges of socio-technical systems in the current dynamic, complex USAF environment. Recently, she has turned her attention to the issues of teams working with small/micro-unmanned air vehicles. Dr Miller co-edited a book providing an overview of the theoretical foundations of macrocognition and macrocognitive metrics to evaluate and measure the performance of real-world teams. Additionally, she has published numerous journal articles and conference papers in areas of human factors.

**Alexander Nelson** is currently pursuing a Masters degree in Cyber Warfare from the Center of Cyberspace Research (CCR), Air Force Institute of Technology (AFIT) where he is studying the psychological foundations of influence as they apply in the cyber domain. Prior to this, he spent six years as a Research Psychologist and Acquisition Program Manager at the Air Force Research Laboratory's (AFRL) Human Effectiveness Directorate, Wright-Patterson AFB, Ohio. Before beginning his service with AFRL in 2004, Mr. Nelson worked as a Research Assistant in the Department of Psychology, Wright State University conducting research on various team training protocols and display design in addition to pursuing graduate coursework in human factors engineering. Mr. Nelson obtained a BA in Psychology in 2001.

**Heather Neyedli** is a PhD candidate in the Department of Exercise Science at the University of Toronto where she is interested in examining the neural and behavioural interactions between attention, action planning and decision making. She completed her Masters of Applied Science in the Mechanical and Industrial Engineering Department at the University of Toronto in 2009 and a BS in Kinesiology (Hons) with Distinction at Dalhousie University in 2007 where she was awarded the University Medal for graduating at the top of the class in the Faculty of Health and Human Performance, as well as the Canadian Society of Exercise Physiology Science Award. Ms. Neyedli received a Natural Sciences and Engineering Research Canadian Graduate Scholarship for both her masters and doctoral studies. She also received the Defence Research & Development Canada Postgraduate Scholarship in 2007.

**Linda G. Pierce** is a Personnel Research Psychologist at the Federal Aviation Administration (FAA) Civil Aerospace Medical Institute. Dr Pierce's research is in selection of air traffic controllers and organizational design in aviation maintenance. Prior to joining the FAA in 2010, Dr Pierce was a Military Research Psychologist for the US Army. During her 20-year career in military research, Dr Pierce held multiple managerial assignments, while maintaining an active research programme in team and human-system collaboration. Her research influenced the development of military systems designed to support and train nationally and organizationally diverse teams in missions ranging from war fighting to peacekeeping. In her last assignment with the US Army, she established a new research unit in organizational performance to work as part of a multidisciplinary team of experts at Aberdeen Proving Ground to address military challenges in network-enabled operations. Her research has been documented in more than 50 journal articles, book chapters, and technical publications, and she is on the editorial board of *The Journal of the Human Factors and Ergonomics Society*. Dr Pierce earned her doctorate in Industrial and Organizational Psychology from Texas Tech University.

**Eduardo Salas** is Trustee Chair and Professor of Psychology at the University of Central Florida where he also holds an appointment as Program Director for the Human Systems Integration Research Department at the Institute for Simulation and Training. Previously, he was the Director of UCF's Applied Experimental and Human Factors PhD Programme. Before joining IST, he was a Senior Research Psychologist and Head of the Training Technology Development Branch of NAWC-TSD for 15 years. During this period, Dr Salas served as a Principal Investigator for numerous R&D programs which focused on teamwork, team training, decision making under stress, and performance assessment. Dr Salas has co-authored over 375 journal articles and book chapters and has co-edited 24 books. Dr Salas is President of the Society for Industrial and Organizational Psychologists, and a Fellow of the American Psychological Association and the Human Factors and Ergonomics Society.

**Paul Salmon** is a Senior Research Fellow within the Human Factors Group at the Monash University Accident Research Centre and currently holds an Australian National Health and Medical Research Council Postdoctoral Training Fellowship in the area of public health. Paul has almost a decade worth of experience of applied human factors research in a number of domains, including the military, aviation, rail and road transport. This has led to Paul gaining expertise in a broad range of areas, including situation awareness, human error and the application of human factors methods, including human error identification, situation awareness measurement, teamwork assessment, task analysis, and cognitive task analysis methods. Paul has authored and co-authored eight books and numerous peer-reviewed journal articles, conference articles, and book chapters, and was recently awarded the 2007 Royal Aeronautical Society Hodgson Prize for a co-authored paper in the society's *Aeronautical Journal*, and, along with his colleagues from the Human Factors Integration Defence Technology Centre (HFI DTC) consortium, the Ergonomics Society's President's Medal in 2008.

**Tamera R. Schneider** is an Associate Professor of Psychology at Wright State University. She earned a PhD in Social/Health Psychology with a focus in Psychophysiology and a minor in Quantitative Methods in 1997 from the State University of New York at Stony Brook. Prior to joining WSU in 2000, Dr Schneider served as a Postdoctoral Fellow at Yale University. She has been awarded numerous teaching and research awards, including the Presidential Award for Faculty Excellence in Early Career Achievement in 2004. Dr Schneider has over 30 publications and 75 presentations. While at WSU, Dr Schneider has acquired over $300,000 in extramural funding as PI and $3.9 million as Co-PI. In addition to funding on organizational climate and faculty well-being, Dr Schneider is currently funded to expand upon her work on stress and emotion as they relate to discerning deception and imaging of facial expressions. Dr Schneider's research interests include psychological and physiological stress resilience, and persuasion for the purpose of promoting health and well-being.

**Gabriella Severe** is a Research Psychologist in the Training and Human Performance Research & Development Branch at the Naval Air Warfare Center, Training Systems Division, in Orlando. Ms. Severe received an MS degree in Industrial and Organizational Psychology and her BS degree in Psychology from the University of Central Florida. She is currently working in the areas of: command and control assessment tools, culture, and trust in the military. Her work has involved collaborations with a variety of organizations such as: Defense Equal Opportunity Management Institute (DEOMI), Deference Language Office (DLO), and North Atlantic Treaty Organization (NATO). Ms. Severe's research interests include the standardization of training evaluation for cultural awareness courses.

**Neville Stanton** was appointed to a Chair in Human Factors in the School of Civil Engineering and the Environment at the University of Southampton in February 2009. Prior to that he held a Chair in Human Factors at Brunel University (since September 1999). Before that he held a Lectureship then Readership in Engineering Psychology at the University of Southampton (since September 1993). He has published over 150 peer-reviewed journal papers (including papers in *Nature* and *New Scientist*) and 20 books on human factors and ergonomics. In 1998 he was awarded the Institution of Electrical Engineers Divisional Premium Award for a co-authored paper on Engineering Psychology and System Safety. The Ergonomics Society awarded him the Otto Edholm medal in 2001 and The President's Medal in 2008 for his contribution to basic and applied ergonomics research. In 2007 The Royal Aeronautical Society awarded him the Hodgson Medal and Bronze Award with colleagues for their work on flight-deck safety. Professor Stanton is an editor of the journal *Ergonomics* and on the editorial board of *Theoretical Issues* *in Ergonomics Science*. Professor Stanton is a Fellow and Chartered Occupational Psychologist registered with The British Psychological Society, and a Fellow of The Ergonomics Society. He has a BSc (Hons) in Occupational Psychology from the University of Hull and a PhD in Human Factors from Aston University in Birmingham.

**Charlene K. Stokes** is a Research Psychologist for the Air Force Research Laboratory, Sensemaking and Organizational Effectiveness Branch at Wright Patterson Air Force Base in Dayton, Ohio (since June 2009). Prior to that, she completed a Postdoctoral Research Fellowship with the same group. She has a BS (Hons) in Psychology from The Ohio State University and an MS and PhD in Industrial/Organizational Psychology from Wright State University. Dr Stokes leads a research programme that includes the examination of trust, emotions, and the development and testing of performance models, with a specific interest in adaptive performance. Her other research interests include cultural adaptability, organizational effectiveness, and stress and coping.

**Megan Thompson** obtained her Masters degree and doctorate in Social Psychology from the University of Waterloo. She began her DND career as an National Sciences and Engineering Research Council Visiting Fellow (NSERC) at the Defence and Civil Institute of Environmental Medicine (DCIEM), subsequently becoming a Defence Scientist at Defence Research & Development Canada (DRDC) in Toronto. Dr Thompson is currently Section Senior Scientist of the Collaborative Learning and Performance Section (CPL) of DRDC Toronto, where her principal research foci involve optimizing moral and ethical decision making in military operations and collaboration within diverse teams, with a focus on optimizing trust. Prior to this she led the Stress and Coping Group for five years, developing a new research programme within DRDC Toronto in the area of psychological adaptation and resiliency in deployments. Dr Thompson has served on several international defence research panels (NATO-179 – Moral Dilemmas and Military Mental Health Outcomes; National Lead, TTCP TP-10 Survival Psychology, National Lead TP-13 Psychological Health and Operational Effectiveness) and was a keynote speaker at the NATO HFM-142 Symposium on 'Adaptability in Coalition Teamwork'. She has authored 41 publications in the areas of team effectiveness, individual differences in judgment and decision making, deployment stress and resiliency, and moral and ethical decision making in operations. She is on the editorial board of the American Psychological Association journal *Military Psychology*.

**Guy Walker** read for a BSc (Hons) degree in Psychology at Southampton University specialising in engineering psychology, statistics and psychophysics. During his undergraduate studies, he also undertook work in auditory perception laboratories at Essex University and the Applied Psychology Unit at Cambridge University. After graduating in 1999, he moved to Brunel University, gaining a PhD in Human Factors in 2002. His research focused on driver performance, situational awareness, and the role of feedback in vehicles. Since this time, he has worked for a human factors consultancy on a project funded by the Rail Safety and Standards Board, examining driver behaviour in relation to warning systems and alarms fitted in train cabs. Currently Dr Walker works within the HFI DTC at the University of Southampton, engaged primarily in work on future command and control systems. He is also author of numerous journal articles and book contributions, and along with his colleagues on the HFI DTC was recently awarded the Ergonomics Society's President's Medal.

**Jessica L. Wildman** is a doctoral candidate in the Industrial and Organizational Psychology Programme at the University of Central Florida, where she earned a BS in Psychology in 2007 and an MS in Industrial and Organizational Psychology in 2010. Since 2007, she has been working closely with her mentor and advisor, Dr Eduardo Salas, on a variety of projects related to culture, teams, trust, and performance as a Graduate Research Associate at the Institute for Simulation and Training. In particular, she was involved in a project funded by the Army Research Laboratory examining the impact of culture on team performance, she was the Student Scientific Lead on a project examining the development of trust in swift starting team contexts, and is currently a Fellow and Student Lead on a MURI grant focused on examining the effects of culture on collaboration and negotiation. She has co-authored six book chapters and four published or in press peer-reviewed journal articles and has personally presented 15 papers or posters at professional conferences. Her current research interests include multicultural performance, team process and performance, and interpersonal trust development and repair.

# Chapter 1
# An Introduction to Trust in Military Teams

Neville A. Stanton
*School of Civil Engineering and the Environment, University of Southampton, Southampton, UK*

## An Introduction to Trust in Military Teams

Trust is something of a nebulous concept; whilst most would agree that establishing trust in teamworking is important (essential even), it can be difficult to define and measure. The *Oxford English Dictionary* defines trust as the *'firm belief in the reliability, truth, or ability of someone or something'* but this definition does not seem to lend itself easily to the concept of trust within and between teams, as the focus of the definition is upon the singular person or artefact. Ashleigh and Stanton (2001) point out that there is no single, universal definition of trust that everyone agrees upon. This may reflect the state of the art, that trust is a concept in development, and the myriad of different situations and domains where trust is being investigated. Perhaps there can be no single overarching definition that would suit every eventuality, situation or domain. Nevertheless, investigations into trust have been a significant challenge on the research agenda for decades. In this book, trust is examined at various levels, including interpersonal, inter-team (i.e., between teams), intra-team (i.e., within teams), inter-organisation (i.e., between organisations) and intra-organisational (i.e., within organisations), as well as trust in technology. This variety puts strain on any theoretical and measurable concept. To return to the work of Ashleigh and Stanton (2001), they identified nine constructs that appeared to be common to intra-team, inter-team and technological trust in a study based on the human supervisory control domain (which bears some similarities to command and control: Stanton, Baber and Harris, 2008). These factors are shown in Table 1.1, with rankings of the importance of each construct for intra-team, inter-team and technology trust.

**Table 1.1 Rankings of trust constructs**

| Construct | Intra-team | Inter-team | Technology |
|---|---|---|---|
| Honesty | 1 | 3 | 6 |
| Understanding | 2 | 2 | 4 |
| Respect | 3 | 5 | 6 |
| Quality of interaction | 4 | 1 | 1 |
| Reliability | 5 | 4 | 2 |
| Communication | 5 | 4 | 4 |
| Ability | 6 | 5 | 5 |
| Performance | 6 | 6 | 3 |
| Expectancy | 7 | 6 | 4 |

The rankings in Table 1.1 show that the inter-team rakings are similar to the intra-team rankings apart from the 'quality of interaction' and 'respect' constructs, where they are similar to the technology constructs. The constructs in this table show that trust is multi-dimensional, and the importance of each construct may change depending on the situation and domain under investigation. It is likely that the rankings of these constructs would be different for interpersonal, inter-organisation and intra-organisational trust. One of the strengths, or weaknesses, of the trust domain is the reliance upon self-report measures. It is a strength by virtue of the fact that the reports are an indication of how a situation is perceived at that moment in time, but it suffers from all the normal drawbacks of subjective assessments. Annett (2002) identifies four key issues for subjective rating scales: subjectivity, measurement, multiple dimensions and fitness. First, he argues that subjectivity is all-pervasive. Even so-called objective methods are inevitably influenced by subjectivity. Second, he argues that subjective judgement can quite successfully be mapped onto ordinal scales for the purpose of measurement. This offers the analyst a means of collecting data about a phenomenon that might not otherwise be available. Third, he argues that it is possible to have both physical and subjective components of multi-dimensional constructs. The critical issue is to establish their validity. There has been much research on the need to establish reliability and validity of Human Factors methods in recent years (e.g., Stanton et al., 2005a, b). Finally, he argues that care should be taken to make sure that a subjective rating scale is fit-for-purpose. Rather than selecting a technique because of its ease of administration, researchers need to determine that the approach is appropriate and will yield useful data for the study they are undertaking. Each of these issues is just as pertinent to subjective measures of trust as to any other subjective assessment.

This book is arranged in three sections. The first section presents chapters on experimental studies into mood, personality and training. In the second section, chapters on multinational challenges for trust are presented. Chapters on trust in technology are presented in the third section.

## Experimental Studies into Mood, Personality and Training

This section focuses on trust at the team level. It contains three chapters: Stokes, Lyons and Schneider (Chapter 2: The impact of mood on interpersonal trust), Lyons, Stokes and Schneider (Chapter 3: Predictors and outcomes of trust in teams) and Lyons et al. (Chapter 4: Exploring the impact of cross-training on team process).

In Chapter 2, Stokes et al. note that computer-mediated communication and coordination are increasingly prevalent in military systems. These systems may be putting more stress on trust in multinational operations if there is a commensurate reduction in face-to-face contact. Trust may be viewed as an essential component for effective cooperation and performance of teams. Stokes et al. argue that the boundaries between affective and cognitive states are somewhat blurred, as trust appears to occupy both types of state. Mood states seem to have an effect on trust, which suggests that affective states may affect cognitive states and vice versa, there being permeable boundaries between both states. This effect appears to diminish over time as cognitive states develop evidence from experience. The authors report on a study to investigate whether positive mood is related to higher levels and negative mood is related to lower levels of team trust. Small teams of five participants undertook an interdependent, computer-mediated task which required a high degree of coordination and communication via instant messaging. The results of the study provide evidence to support the hypothesised link between mood and trust, both at the individual and team level. This suggests that a relatively simple way of increasing team trust would be to manage the mood of the team.

Lyons et al., in Chapter 3, investigate the reasons why face-to-face teams have higher levels of trust than distributed teams. Part of the reason for the high level of trust is simply exposure. Distributed teams are likely to interact less than co-located teams, as there will be more opportunities for interactions when teams are face-to-face, and these interactions are likely to be 'richer', including non-work (social) interactions. There are also bandwidth restrictions for distributed teams, which often reduce the fidelity of non-verbal communication. An analysis of the relationship between personality characteristics and trust led to the identification of positive links between 'agreeableness', 'extroversion' and 'openness' and trust, as well as potential negative links between 'neuroticism' and trust. Small teams of five participants took part in a computer-mediated task communicating via instant messaging. Measures were taken of cohesion, cognitive ability, personality, interpersonal trust and task performance. The results of the study showed that 'agreeableness' and 'openness' did predict trust. The results did not show any links between 'extroversion' and neuroticism' and trust. The authors suggest that the computer-mediated nature of the task did not allow these latter two personality characteristics to exhibit themselves and therefore negated any effect that they may have had.

In Chapter 4, Lyons et al. argue that cross-team training can help the team develop as a functionally interdependent unit, establishing common ground, trust and cohesion through coordination of activity and development of shared mental models. Team process variables such as communication, interaction, cohesion and trust are developed in cross-training activities. Trust, in particular, is dynamic in nature and open to change during cross-training as people gain experience of working together. Positive aspects of trust within teams have significant benefits, such as reducing the team process overheads, increasing collective responsibility and encouraging sharing behaviour. Lyons et al. conducted studies with small five-person teams in a simulated command and control task over five days (with members of the team either performing the same task over the five consecutive days in the 'permanent' condition or a different task in the 'rotation' condition). It was hypothesised that people in the 'rotation' condition would report higher levels of trust, as the rotational cross-training would promote more common ground and shared understanding of each other's roles. Measures were taken of team performance, communications, cohesion, trust and collective efficacy. The finding worth highlighting is that, contrary to expectations, 'permanent' teams reported a higher level of trust than the 'rotation' teams. The authors speculate that the rotation of the roles may have been more disruptive to team processes than first anticipated, as rotation effectively created a new team structure each time the roles changed. This change may have inhibited team development.

## Multinational Challenges for Trust

There are four chapters on the multinational challenges for trust: Wildman et al. (Chapter 5: Trust in swift starting action teams), Cianciolo et al. (Chapter 6: Trust in distributed operations), Gill, Thompson and Febrarro (Chapter 7: Trust in international military missions) and Hughes et al. (Chapter 8: Cultural influences on trust).

In Chapter 5, Wildman et al. discuss the role of swift starting action teams in multinational collaborative operations. The idea behind swift starting action teams is that they are able to develop appropriate levels of trust quickly to enable them to function at the optimum level in their performance of teamwork in the shortest possible time. By their very nature, swift starting action teams do not have the luxury of time to build up trusting relationships in the normal manner. Inappropriate levels of trust may lead to process inefficiencies (e.g., wasted effort checking another party's work or failure to identify inaction by another person due to a failure to monitor them appropriately). Rapid calibration of swift starting action teams is required but most models of interpersonal trust assume a period of trust building before trust is calibrated. By contrast, 'swift trust' proposes that trust can be transferred from one domain to another. Swift trust would seem to rely upon trust schemata, developed through experience in other domains that appears to suit the current situation. These schemata appear to offer possible 'trust templates' for immediate initial calibration of trust in team members that can be modified through experience and stored for new situations. One might assume that swift starting action teams become more adept at selecting the appropriate 'trust template' as they gain more experience of swift team operations. Wildman et al. are keen to point out that swift trust may be mis-calibrated, which could be detrimental to team performance. If swift teams only come together for brief periods, any mis-calibration might not be detected for the time the team is together. The authors propose a range of training strategies that could help mitigate against trust biases in swift starting action teams.

Cianciolo et al. examine the nature of interpersonal trust in Chapter 6, within the context of network enabled command and control. New systems and distributed operations pose considerable challenges for trust, and behaviours and attitudes may be mediated by computerised communication technology. This complexity is further exacerbated by the context of joint, interagency, intergovernmental, multinational operations. Such operations are often characterised by ill-defined problems, distributed decision making and blended communications (i.e., a mixture of face-to-face and technology-mediated exchanges). The authors proposed a model of trust for distributed operations to represent the main factors that mediate between trustor and trustee behaviour. The model proposed that trustworthiness of the trustee can be determined by competence, character and dependability. This in turn leads to trustee behaviours which interact with the trustor characteristics, which ultimately lead to a trustworthiness judgement. These are moderated by situational and personal characteristics to produce the trust related behaviours. The bias towards positive or negative expectations will affect the degree of risk management associated with the trust related behaviours the trustor engages in. Cianciolo et al. propose three levels of analysis for investigations of trust in military teams that best characterise the diverse set of interactions: 'co-located', 'distributed and interoperable' and 'distributed and non-interoperable'.

Gill et al. consider trust as an essential dimension in modern operations in Chapter 7. The presence of trust has positive connotations for the complexities of contemporary operations (which range from warfighting through reconstruction to stabilisation), but violations of trust can potentially be very destructive. The effects of the violations of trust seem to depend upon the nature of the violations and the way in which the repair strategy is implemented. Anomalous violations are easier to salvage than those considered to reveal a lack of integrity. In a military setting, an example of an anomalous violation might be inflicting collateral damage when none had occurred in operations before, whereas an example of a violation of integrity might be failing to deliver on sincere promises to restore facilities and utilities. In order to explore the relationship between violations and repair strategy, Gill et al. conducted a controlled study by manipulating the independent variables of trust violation (i.e., lack of competence versus lack of integrity) and repair strategy (i.e., apology versus denial). They found that trust decreased after both types of violation on self-report measures. Trust was recovered to some extent after both types of repair strategy (in contrast with previous research which favoured the denial to the apology), although the apology appeared to be the best repair strategy in terms of improving trust. The authors explain this contradiction in terms of the different domains and context of the research compared to previous research on repairing trust.

In Chapter 8, Hughes et al. consider cross-cultural competencies. They argue that these have become an important focus now that armed forces are primarily engaged in stability operations. Military personnel are increasingly required to interact with civilian populations, and trust is an essential factor in developing good relationships with local people. Hughes et al. sought to extend models of trust by investigating the relationship between cross-cultural competence and trust. They identified potential impacts that cultural distance and biases might have on trust. Cultural distance is defined by the magnitude of differences between cultures. The greater the degree of differences between cultures, the greater the difficulty in establishing trust between those cultures. They based this observation on reports of shared experience, common frames of reference, expectations and potential communication difficulties. Cultural biases can be rooted in particular ideologies, cultural norms and values that are barriers to the development of interpersonal trust. Understanding these cultural differences and biases can help overcome the barriers to trust, and adopting the other person's conventions can help in the development of trust.

## Trust in Technology

Four of the chapters deal with the effects that technology may have upon trust: Hollands and Neyedi (Chapter 9: A reliance model for automated combat identification systems), Salmon et al. (Chapter 10: User trust in new battle management technology), Fitzhugh, Hoffman and Miller (Chapter 11: Active trust management) and Masys (Chapter 12: Teamwork and trust: a socio-technical perspective).

In chapter 9, Hollands and Neyedi consider the level of trust in the combat identification system that operators develop. Although combat identification systems offer the potential to detect friendly assets and are highly unlikely to present a false positive (i.e., report a friend as an enemy) they are prone to presenting false negatives (i.e., reporting a friendly, neutral or enemy asset as an 'unknown'). Hollands and Neyedi argue that trust is not just an interpersonal phenomenon; it also applies to human–machine interactions and may share some of the same characteristics. As with interpersonal trust, trust in technology and automation needs to be appropriate to the capabilities of those technologies, as the level of trust will determine the nature of the human interaction and the resultant system performance. The authors point out that there are some important differences between interpersonal trust and trust in automation. They suggest that trust is more fragile and the reliability of the system has implications for the way in which it will be used. Research into varying levels of reliability of a combat identification system showed that operators changed their level of reliance on the system only when information about reliability was communicated. This suggests that the system operators were not able to gauge the level of reliability themselves. This finding was contradicted by another study which revealed that under some circumstances people are able to judge system reliability.

In Chapter 10, Salmon et al. investigate trust in the context of new technology. There is increasing reliance on computers in mission planning and battle-space management, which puts an increased emphasis on trust in technology to deliver effective performance. This aspiration becomes increasingly pertinent with rapid tempo operations placing demand on technology to support rapid planning and up-to-the-minute accuracy on the position tracking of assets and their status. Salmon et al. explore the relationship between trust and situation awareness in the context of an evaluation of a new digital mission planning and battle-space management system. The system was intended to replace traditional paper-based mission planning and verbal situation reports via radio for battle-space management. They observed a headquarters staff perform their tasks during a three-week operational field trial. Analysis of the situation awareness data revealed that the staff officers mistrusted much of the data presented to them on the battle-space management system, such as the position of blue (i.e., friendly) and red (i.e., enemy) forces, because of the lack of timeliness of the information. Contact reports were up to twenty minutes out of date. In addition, some of this information was spurious, either under- or over-reporting on enemy strength. The authors argue that low levels of trust in the information had an adverse effect on the acquisition and maintenance of situation awareness in the headquarters, and therefore a reluctance to accept the new technology.

Fitzhugh, Hoffman and Miller explore active trust management in Chapter 11. Cyber systems are vulnerable to attack and corruption, and therefore require monitoring and checking. Active trust management implies continuous assessment of trust in technological systems. The authors propose a trust decision cycle which emphasises the test and evaluation of trust in measurable terms. They argue that trust is a dynamic concept, moving along a continuum as evidence for and against trust a system is gathered and verified. Rather than presenting a black-and-white picture of trust, of absolute trust or absolute mistrust, this is a view of trust as shades of grey moving up and down the spectrum. Cyber systems are under constant attack from a range of sources, including enemy states, terror organisations, criminals and hackers. Fitzhugh, Hoffman and Miller identify four facets of trust that are pertinent to the cyber domain: trust in technology (e.g., data integrity, transmission integrity, source credibility, system opacity and system reliability), trust in decision making (e.g., compromised decision information storage, transmission or processing), trust in the socio-technical system (e.g., risk, uncertainty, reliability, accessibility and accountability) and trust in overall capability (e.g., measures of output such as usefulness, completeness and timeliness). They demonstrate the application of these concepts to a UAV Predator system to show the relevance of their concerns.

Finally, in Chapter 12, Masys takes an Actor Network Theory perspective to investigate trust in military socio-technical systems. Actor Network Theory does not make a distinction between the social and technical actors. Both types of actors have competencies which contribute in combination to produce system performance. Fundamentally, Actor Network Theory seeks to understand how the actors' joint patterns of behaviour, relations, interactions and associations change and transform over space and time and the subsequent impact on the system. Transformations in the networks may reveal the power of actors or cliques have to change other actors. Trust has an important role to play in the behaviour of the actors and hence the networks. Human–human trust as well as trust mediated by, and trust in, technology is likely to be affected (both positively and negatively) through actor interactions. In case studies of fratricide incidents, Masys shows how human relations with, and trust in, technology is similar to human–human relations and trust. The analysis of socio-technical systems as actor networks reveals the fundamental nature of systems that only makes sense in terms of the 'hybrid collectif' comprising of people + technology + mission + training + standard operating procedures + doctrine + culture, and so on. It is these complex interactions and interdependencies that produce the patterns of systems behaviour and transformations. To paraphrase Masys, the web of interrelations and interactions give rise to trust in systems, and the analysis should remain at the level of the system rather than the individual entities. That is not to say that trust will necessarily be appropriate, as socio-technical systems can have a misplaced sense of trust or distrust in the actors and their competencies, such as the reliability and validity of Identification of Friend or Foe systems.

## Conclusions for Trust in Military Teams

In summary, this book covers a broad spectrum of issues in the 10 chapters that follow, from technology to the individual, to the team to the organisation to national collaborations. As mentioned earlier, the facets of trust vary with the level and focus of the analysis. Despite this variety, or perhaps because of it, some general practical conclusions can be drawn from the book, as follows:

- positive mood is related to higher levels and negative mood is related to lower levels of team trust;
- personality characteristics of 'agreeableness' and 'openness' can be used to predict trust;
- permanent teams report higher levels of trust than non-permanent teams;
- trust templates may be used for initial calibration of trust in ad-hoc teams;

- the interaction of trustor and trustee behaviour leads to the trustworthiness judgement;
- apology is the best trust repair strategy;
- adopting the other person's conventions can help in the development of trust;
- under certain situations people can develop appropriate levels of trust in technology;
- low levels of trust in technology have an adverse affect upon the acquisition and maintenance of situation awareness;
- technological systems can require continuous assessment and verification of trust;
- the trust system is the integration of people, technology, procedures, training, doctrine and culture, and needs to be studied as a whole.

The contributing chapters to this book also show that much more research into trust is needed, particularly into technology, teams, organisations, and multinational aspects of trust. The research directions for each of these areas are indicated within the individual chapters.

## References

Annett, J. (2002). Subjective rating scales: science or art? *Ergonomics*, 45(14), 966–87.

Ashleigh M.J. and Stanton N.A. (2001). Trust, key elements in human supervisory control domains. *Cognition, Work and Technology*, 3(2), 92–100.

Stanton N.A., Hedge A., Salas E., Hendrick H. and Brookhuis K. (2005a). *Handbook of Human Factors and Ergonomics Methods.* CRC Press, Taylor & Francis, London, UK.

Stanton N.A., Salmon P.M., Walker G.H., Baber C. and Jenkins D. (2005b). *Human Factors Methods: A Practical Guide for Engineering and Design.* Ashgate Publishing Ltd, Aldershot, UK.

Stanton N.A., Baber C. and Harris D. (2008). *Modelling Command and Control: Event Analysis of Systemic Teamwork.* Ashgate Publishing Ltd, Aldershot, UK.

# SECTION ONE
# Experimental Studies into Mood, Personality and Training

# Chapter 2
# The Impact of Mood on Interpersonal Trust: Implications for Multicultural Teams

Charlene K. Stokes and Joseph B. Lyons
*US Air Force Research Laboratory, Dayton, OH, USA*

Tamera R. Schneider
*Wright State University, Dayton, OH, USA*

## Introduction

Virtual teams that coordinate and communicate remotely using computer-mediated mechanisms are becoming prolific in military and other organizational settings today (see Lyons, Stokes and Schneider, this volume). Many of these settings are characterized by teams with members from differing cultural backgrounds. For example, US military adversarial and humanitarian missions are increasingly shifting toward multinational operations (Klein and Steele-Johnson, 2007). A growing body of research highlights the critical role trust plays in enhancing the coordination and performance of such teams (Aubert and Kelsey, 2003; Dirks, 1999; Fuller, Mayer, and Pike, 2008; Jarvenpaa and Leidner, 1999; Jones and George, 1998; Lyons, Stokes, and Schneider, this volume). In an examination of the effects of interpersonal trust on team performance, Dirks (1999) found that trust increases team performance through improvements in team effectiveness and efficiency, maximizing use of team resources. Trust improves team effectiveness by increasing the motivation of team members to cooperate and work together, and trust improves team efficiency through reduced need for controls (e.g., rules, monitoring behavior) and increased ability to confront performance problems (Dirks, 1999). Thus, developing an understanding of ways to increase interpersonal or team trust in virtual environments offers a key leverage point for increasing team coordination and performance.

### *Trust Defined*

Trust has been defined as the acceptance of risk or vulnerability from others based on positive expectations (Mayer, Davis, and Schoorman, 1995). Trust is therefore construed as an attitude that impacts judgment and behavior. It is important to distinguish the outcome of trust from its two primary antecedents: trustworthiness and trust propensity. Trustworthiness is defined as characteristics associated with a target (i.e., the trustee) that inspire positive expectations, and trust propensity is a dispositional willingness to rely on others (Colquitt, Scott, and LePine,

2007). Trust can be modified in two primary and interrelated ways: 1) through trustworthiness characteristics of the trustee (human or machine) or 2) through perceptions and characteristics of the trustor, such as trust propensity or situational aspects. If focusing on aspects of the trustee, enhancing the key trustworthiness indicators of ability, integrity, or benevolence will increase interpersonal trust (Mayer et al., 1995). To increase trust in technology, a set of factors closely related to the indicators described by Mayer et al. (1995) but applied to technology are enhancement of machine competence, predictability, dependability, and transparency (Muir, 1994). However, these trustworthiness indicators must be perceived by the trustor for a precise calibration of trust in another person or in the technology being used. As with perception of any environmental stimuli, there are numerous psychological, contextual (e.g., culture), and physiological factors that can impact the perception of trustworthiness and trust judgments. For example, Lyons and colleagues (this volume) provide empirical evidence for the impact of the dispositional factors of agreeableness, openness, and cognitive ability on trust. However, factors as seemingly innocuous as an individual's affective state can have a significant impact on perception and judgment, and therefore trust.

### *Mood and Trust*

Moods and emotions are affective states or feelings that inform individuals of their general state of being as they go about their day and interact with others (Nowlis, 1970, as described in Jones and George, 1998). Emotions generally have a specific cause and are usually more intense in experience and short in duration (seconds to minutes), whereas mood tends to be less intense and longer lasting, capturing more of the stable day-to-day feeling that people experience (Forgas, 1995; Jones and George, 1998; Zimmermann et al. 2003). Although many authors use the term emotion as all-encompassing, the majority of affect-based research assesses the more subtle affective variable of mood, and as such the term *mood* will primarily be used in the present paper. The generic term *affect* does encompass moods and emotions (Forgas, 1995) and is often differentiated by time and degree of impact. *State affect* is a temporary state of feeling and *trait affect* is a dispositional tendency to experience relatively more positive or negative moods and emotions (Gaudry, Vagg, and Spielberger, 1975; Watson, Clark, and Tellegen, 1988). From an evolutionary standpoint, emotions contribute to motivational systems that have evolved to mediate transactions in the environment that either promote or threaten survival (Lang, Bradley, and Cuthbert, 1990; Lazarus, 1999). Trust, which hinges on judgments of vulnerability in an effort to reduce risk, is likely driven by similar motivational mechanisms. Given that emotions and trust appear to share a similar evolutionary purpose, it is plausible to expect a relationship between the two. Even though moods are more subtle than emotions, research has shown they can have a significant impact on behavior (Jones and George, 1998), and although few have empirically tested the assertion, numerous researchers have postulated a direct relationship between mood and trust.

Mood impacts trust development through cognitive, motivational, and behavioral sources (Williams, 2001). Cognitively, mood impacts the trust perceptions of others by directing attention and influencing how trustworthiness indicators are evaluated and judged (Forgas, 1995; Jones and George, 1998). Motivationally, mood impacts how willing one is to trust in others (Schoorman, Mayer, and Davis, 2007; Williams, 2001, 2007), and behaviorally, mood impacts how inclined one is to cooperate with others (Webber, 2008; Williams, 2001, 2007) and take action based on another's recommendation (Stokes and Lyons, 2009). Taken together, as these three sources are not mutually exclusive, moods provide individuals with information on how they are experiencing trust at a fundamental level (Jones and George, 1998).

*Mood effects on type of trust* There has been debate regarding the boundary conditions for the impact of mood on trust. Affect is posited to differentially relate to trust across type and time. In terms of type of trust impacted, McAllister (1995) and Rempel, Holmes, and Zanna (1985) asserted that mood primarily influences the deeper stages of trust, which McAllister referred to as affect-based trust. McAllister (1995) distinguished between affect- and cognitive-based trust, with the former being harder to break, longer to establish, based on emotional bonds, and more stable over time and situations. In contrast, cognitive-based trust is more superficial and derived from rational beliefs about reliability, dependability, and competence (McAllister, 1995). Jones and George (1998) presented a similar distinction in unconditional trust, based on positive affect and mutual identification, and conditional trust, based on knowledge and positive expectations. Affect clearly plays a significant role in the deep emotional bonds inherent in affect-based and unconditional trust, but this does not diminish the important role of affect in other cognitively oriented forms of trust. Tomlinson and Mayer (2009) suggested that, even when the source of mood is distinct from the target being evaluated, moods provide a psychological context in which judgments of trust and trustworthiness are made. Empirical evidence provided by Dunn and Schweitzer (2005) confirmed this assertion. They found that mood states unrelated to the trustee, in this case co-workers, or situation at hand had an effect on trust in co-workers in a particular situation. More recent evidence by Lount (2010) indicated that induced states of positive mood influenced initial trust judgments of new team members. Thus, affect does play a role in cognitively oriented forms of trust, but its influence operates indirectly through its impact on other mechanisms such as information processing.

There are clearly many mechanisms involved that contribute to mood having a more holistic impact on trust. The impact of mood on cognitive processes related to trust (e.g., perceptions, beliefs, and judgments) has been well established. Unlike the traditional view of affect as being irrelevant for cognitive processes, as McAllister's (1995) distinction seems to imply, contemporary theories and models of affect view it as an inseparable component or input for the cognitively driven information processing and retrieval systems. The affect infusion model, or AIM

(Forgas, 1995), integrates two strongly supported social cognitive theories of how affect influences cognitive processes and ultimately impacts judgment in social situations, such as the decision to trust another. AIM proposes that affect operates through the two mechanisms of affect-priming and affect-as-information. The two mechanisms are not mutually exclusive and can occur simultaneously, but the relative strength of each mechanism is determined by the dominant processing strategy adopted. During in-depth, substantive processing activities, such as those involved in social judgments in complex and novel situations, the affect-priming principle is dominant. This principle suggests that affect influences judgments indirectly through its effect on the selective focusing of attention, encoding, retrieval, and associative processes. In contrast, the affect-as-information mechanism is dominant during faster, heuristic processing, where feelings directly influence judgments as individuals use their affective states as a short-cut to infer their evaluative reactions to a target. Forgas (1995) reviews a long history of empirical evidence in support of these two affect infusion mechanisms. In summary, the evidence outlined above suggests that mood not only plays a direct role in affect-based trust, but it also plays a significant role in cognitive-based forms of trust through its impact on various cognitive mediating mechanisms.

*Mood effects on trust over time* Research has shown that the impact of mood on trust diminishes over time (Stokes, Lyons, Littlejohn, Natarian, Case, and Speranza , 2010), which is likely related to changes over time in the trust construct itself. Merritt and Ilgen (2008) found that two qualitatively different trust constructs, dispositional and history-based trust, emerge over time and show distinct relationships with other factors – although mood was not one of the factors examined. Dispositional trust, similar to what Mayer and colleagues (1995) termed initial trust, forms early on and is largely dependent on one's trust propensity due to the limited interaction or knowledge of the trustee. History-based trust forms over time following repeated interactions that allow the trustor more opportunities to gauge the trustworthiness of the other (Merritt and Ilgen, 2008). In the absence of the necessary interaction to accurately calibrate trustworthiness, individuals likely rely on numerous other factors such as similarity, group membership, and contextual cues (Kramer, 1999; Lount, 2010; Tanis and Postmes, 2005) to form judgments of trust early on. Indeed, our previous research (Stokes et al., 2010) has found that mood serves as a key factor early on in the trust process, whereas subsequent trust was driven more by interaction with the trustee, an automated system in this case. Going back to the AIM model (Forgas, 1995), it is likely that the affect-as-information and/or the affect-priming principles are operating at this earlier stage to influence trust judgments.

### *The Current Study*

Considering the positive implications that increased trust has in multinational team settings, mood was examined in the current study as a factor influencing judgments of trust in an interdependent team task composed of team members from various cultural and national backgrounds. Although limited evidence exists for the empirical relationship between mood and the specific construct of trust, a sizeable tangential research base in the social psychological literature exists that supports mood effects on risk taking tendencies. In general, it has been found that positive moods increase acceptance of risk, whereas negative moods promote risk aversion (Chou, Lee, and Ho, 2007; Forgas, 1995; Lerner and Keltner, 2000). Under the AIM model (Forgas, 1995), the directionality of these effects likely occur due to biases in cognitive processing and information retrieval, prompting individuals to evaluate situations in a manner consistent with their current mood. For example, individuals in a positive mood are more likely to access information from pre-existing knowledge structures that are associated with positive moods (Chou et al., 2007). As these knowledge structures are associated with positive moods, the outcomes of the past situations tend to be positive, thereby focusing the individual on the potential of positive outcomes from accepting risk in the present situation. Thus, judgment is largely guided by an activated heuristic that positive outcomes are likely. Negative moods, on the other hand, are associated with increased scrutiny and detailed processing of possible flaws in a situation given the association of flaws in previous situations. Investigating the ability to detect truthfulness in others, Forgas and East (2008) found individuals in negative moods were more skeptical and detected deception more often than individuals in positive moods, whom Forgas and East suggested were more trusting – although a direct measure of trust was not used. Thus, given that accepting risk is commensurate with accepting vulnerability (i.e., trust), the following hypotheses were posited:

> Hypothesis 1: Positive individual mood will be related to higher levels of team trust.

> Hypothesis 2: Negative individual mood will be related to lower levels of team trust.

*Team mood* Although some research and speculation exists regarding the relationship between individual mood and interpersonal trust, no attention has been given to the possible relationship between team mood and interpersonal trust, to our knowledge. Given that interpersonal trust is inherently a relational construct, it is important to consider not only how an individual's mood influences trust but also how the emergence of team-level affect influences trust. As such, perceptions of team mood as a predictor of interpersonal trust were examined in the present study. Team mood as defined herein is not the same as the group-level phenomenon of team affective tone referred to by George (1990), which is the aggregation to the group level of analysis of individually reported affective states, and therefore still based on self-oriented perceptions. Rather, team mood was defined and measured in the present study as self-perceptions of the team's affective tone. That is, as opposed to participants reporting how they personally feel at the moment and then aggregating those results, participants were asked to adopt a team orientation and report what they felt the general mood of the team was during the task, essentially performing their own self-construed aggregation.

Previous research has found consistency between individual mood and team mood, suggesting generalization of mood within a team (George, 1990; Ilies, Wagner, and Morgeson, 2007). Building from Kelly and Barsade's (2001) model of affective influences in groups, Ilies and colleagues (2007) suggest three mechanisms that explain such affective links among team members: common experiences (e.g., success at a group task), behavioral entrainment (e.g., the coordination of behavior among group members, such as in the present study), and emotional contagion (the process of 'catching' another's emotions; see Hatfield, Cacioppo, and Rapson, 1993). Kelly and Barasade (2001) describe such mechanisms as top-down components that create an *affective context* for the group, and the aggregation of individual affect as bottom-up components that create the *affective composition.* These two components then combine to form team mood. Generalized to this team level, George (1990) found that team mood influenced behavioral outcomes in a similar manner as individual mood. Although different constructs, given that team mood and individual mood share similar relationships with certain outcome measures, parallel hypotheses with individual mood were posited for team mood:

> Hypothesis 3: Positive team mood will be related to higher levels of team trust.

> Hypothesis 4: Negative team mood will be related to lower levels of team trust.

## Method

### *Participants*

As part of a larger study, a total of 275 people from a Midwestern university population volunteered to participate in the study in exchange for course credit (200 subjects) or for monetary remuneration in the amount of $30. As the focus of the present research is on multicultural teams, analyses were limited to mixed culture teams only, resulting in a sample size of $N = 42$ teams (210 individuals, 55 per cent female). The majority of the sample was Caucasian (55 per cent) or African American (20 per cent), with each team having at least one member of a different nationality. International students, primarily from India (16 per cent) or other nationalities (9 per cent), comprised the balance of the sample. The age distribution of the sample ranged from 18 to 49, mean = 21.

### *Materials*

*Mood* The positive and negative affect schedule (PANAS; Watson et al., 1988) was used to assess not only individual mood, but the overall perceived mood of the team. Responding on a five-point Likert scale (1 = Very Slightly, 5 = Extremely), participants completed a 10-item version of the PANAS. Five items assessed positive affect (interested, excited, enthusiastic, determined, and alert), and five items assessed negative affect (distressed, upset, irritable, nervous, and jittery). For individual mood, the directions stated '… indicate to what extent *you* feel this way right now,' whereas the directions for team mood stated '… indicate to what extent you feel this represents *your team* right now.' Baseline levels of individual mood were reliably assessed as a control variable (baseline positive mood $\alpha$ = .84, baseline negative mood $\alpha$ = .75), and subsequent measures of individual and team mood were assessed reliably (individual positive mood $\alpha$ = .91, individual negative mood $\alpha$ = .84, team positive mood $\alpha$ = .91, team negative mood $\alpha$ = .86).

*Trust* Trust was measured using a modified version of the Organizational Trust Inventory (OTI; Cummings and Bromiley, 1996). Modified versions of this scale have been used reliably in past research (see Naquin and Paulson, 2003), and the internal consistency for the subset of data analyzed for the present study was high ($\alpha$ = .87). Responding on a seven-point Likert scale (1 = Strongly Disagree, 7 = Strongly Agree) participants rated 11 items on the degree of trust they felt toward their teammates during the task. Example items include, 'I feel that other team members performed the task with me honestly,' and 'I think other team members took advantage of my problems.'

### *Task*

A team-based laboratory task, Computer-based Aerial Port Simulation (CAPS; Lyons, Stokes, Palumbo, Boyle, Seyba, and Ames, 2008), was used as the research platform. The hardware included five networked PCs that participants used to perform the task, and a sixth PC served as the experimenter station for data upload and scenario manipulation. The CAPS software is a computer-generated, five-player simulation program of the logistics operations associated with an aerial port squadron. A team was composed of five functional stations: (a) passenger services, (b) fleet services, (c) cargo services, (d) ramp services, and (d) air terminal operations flight (ATOF). The stations are interdependent; for example, fleet services cannot clean the aircraft until passenger services has unloaded all passengers. Thus, participants must coordinate and communicate their individual activities to achieve the shared goal of preparing aircraft for takeoff in sufficient time. Due to the high degree of communication required to complete this task, a vital component of the CAPS software is the instant message (IM) system, where participants are able to communicate needed information to other team members individually or globally.

*Procedure*

Experimental sessions, lasting approximately 2.5 hours, were composed of a single team of five participants. Each participant was randomly assigned to a task station, where they remained throughout the experimental session, completing all questionnaires and task activities on the computer. After obtaining participants' consent, they completed a battery of pre-task questionnaires that included the baseline assessment of individual mood. After completing the pre-task questionnaires, participants received task instructions and training, followed by two 30-minute task sessions. Measures of individual mood, team mood, and interpersonal trust were collected following the first 30-minute task session.

## Results

Descriptive statistics and correlations are shown in Table 2.1 Consistent with previous research, individual and team mood were related (George, 1990; Ilies et al., 2007), and positive and negative mood were found to be distinct constructs at the individual level (Watson et al., 1988). Interestingly, positive and negative team mood were related, which provides evidence for team mood as a distinct, albeit related construct from individual mood. Bivariate correlations between interpersonal trust and all assessments of mood were in the hypothesized direction: higher levels of trust were related to higher levels of positive mood and to lower levels of negative mood.

**Table 2.1 Descriptive statistics and correlations**

| Variable (*N* = 210) | *M* | *SD* | 1 | 2 | 3 | 4 | 5 | 6 | 7 |
|---|---|---|---|---|---|---|---|---|---|
| Baseline positive mood | 2.9 | 0.8 | *(.84)* | | | | | | |
| Baseline negative mood | 1.7 | 0.7 | .11 | *(.75)* | | | | | |
| Individual positive mood | 3.2 | 1.0 | .58*** | .05 | *(.91)* | | | | |
| Individual negative mood | 1.7 | 0.8 | -.02 | .54*** | .00 | *(.84)* | | | |
| Team positive mood | 4.7 | 1.2 | .37*** | .03 | .59*** | -.07 | *(.91)* | | |
| Team negative mood | 3.4 | 1.3 | -.05 | .29*** | -.08 | .54*** | -.32*** | *(.86)* | |
| Interpersonal trust | 5.4 | 0.9 | .18** | -.29*** | .29*** | -.23** | .52*** | -.43*** | *(.87)* |

*Note:* *p < .05, **p < .01, ***p < .001. Cronbach's Alpha on the diagonal (italicized).

To further explore the relationship between mood and trust, a series of stepwise regression analyses were performed: one examined effects for positive mood variables on trust, and one examined effects for negative mood variables on trust. As displayed in Table 2.2, controlling for baseline positive mood, individual positive mood contributed significantly to the prediction of interpersonal trust, $\Delta R^2 = .05$, $F_{change}$ (1, 207) = 11.55, $p < .01$, supporting Hypothesis 1. Accounting for the greatest amount of unique variance and supporting Hypothesis 3, the addition of team positive mood in Step 3 accounted for incremental variance in interpersonal trust above and beyond that accounted for by individual positive mood, $\Delta R^2 = .18$, $F_{change}$ (1, 206) = 50.88, $p < .001$. As displayed in Table 2.3, individual negative mood was not significantly related to interpersonal trust after controlling for baseline negative mood, which did contribute significantly to the prediction of trust, $R^2 = .08$, $F$ (1, 208) = 19.32, $p < .001$, thus, only partially supporting Hypothesis 2. However, examination of the means for baseline and subsequent individual mood indicated that negative mood did not change during the task (see Table 2.1). Therefore, a follow-up regression analysis was conducted examining individual negative mood without controlling for baseline negative mood. The results indicated that the subsequent measure of individual negative mood was negatively related to trust, as expected: $R^2 = .05$, $F$ (1, 213) = 11.69, $p < .001$. Hypothesis 4 was fully supported in that team negative mood uniquely predicted interpersonal trust, $\Delta R^2 = .13$, $F_{change}$ (1, 206) = 35.90, $p < .001$. Consistent with the relational directions posited in the hypotheses, higher levels of positive mood (individual and team) were related to higher levels of interpersonal trust, and higher levels of negative mood (individual and team) were related to lower levels of interpersonal trust.

**Table 2.2 Hierarchical regression analysis, regressing positive mood variables on interpersonal trust, controlling for baseline individual positive mood**

| Variable ($N$ = 210) | $B$ | $SE\ B$ | Β | $R^2$ | $\Delta R^2$ |
|---|---|---|---|---|---|
| Step 1 | | | | .03 | .03** |
| Baseline individual positive mood | .20 | .07 | .18** | | |
| Step 2 | | | | .08 | .05** |
| Baseline individual positive mood | .02 | .09 | .02 | | |
| Individual positive mood | .25 | .07 | .28** | | |
| Step 3 | | | | .27 | .18*** |
| Baseline individual positive mood | .01 | .08 | .01 | | |
| Individual positive mood | -.03 | .08 | -.04 | | |
| Team positive mood | .40 | .06 | .53*** | | |

**Table 2.3 Hierarchical regression analysis, regressing negative mood variables on interpersonal trust, controlling for baseline individual negative mood**

| **Variable (*N* = 210)** | ***B*** | ***SE B*** | **Β** | ***R*²** | **Δ*R*²** |
|---|---|---|---|---|---|
| Step 1 | | | | .08 | .08*** |
| Baseline individual negative mood | -.36 | .08 | -.29*** | | |
| Step 2 | | | | .09 | .01 |
| Baseline individual negative mood | -.30 | .09 | -.24** | | |
| Individual negative mood | -.11 | .09 | -.09 | | |
| Step 3 | | | | .23 | .13*** |
| Baseline individual negative mood | -.30 | .09 | -.24** | | |
| Individual negative mood | .16 | .09 | .14 | | |
| Team negative mood | -.32 | .05 | -44*** | | |

## Discussion

After nearly three decades of research, trust remains an elusive construct; yet one with tremendous potential to influence human behavior. Until recently, very few studies have explored the link between affective processes and trust, despite an intuitive connection at the motivational/evolutionary level. There is a rich legacy in research exploring the influence of emotions on social and cognitive judgments (Bless, Bohner, Schwarz, and Stack, 1990; Schwarz, 1990; Schwarz and Clore, 1983; Schwarz, Bless, and Bohner, 1991; Williams, 2007). The current study sought to expand this literature by exploring how mood impacts trust in multicultural teams, a relatively untapped area of inquiry to date, though one with considerable implications for military teams. Additionally, mood was explored as an individual and team-level influence on trust. To date, there have been no such studies examining different levels of mood on trust. The implications of this research are discussed below.

### *Individual Mood Effects*

It was hypothesized that individual positive mood would be associated with higher trust. As expected, positive mood was related to higher trust even after controlling for baseline positive mood. This is consistent with research suggesting that positive moods cue benign evaluations of one's environment (Schwarz, 1990) and it supports previous research on mood effects on trust in automation

(Stokes et al., 2010) and interpersonal settings (Lount, 2010). Further, these results support the notion that trust has affective underpinnings, as suggested by several researchers (McAllister, 1995; Schoorman et al., 2007; Williams, 2001, 2007). In contrast, negative individual mood was expected to have a negative association with trust. This hypothesis was supported for baseline levels of individual negative mood. As negative mood did not change during the task, the subsequent measure of individual mood was not significant when controlling for baseline. However, both baseline and subsequent individual mood when analyzed separately had a significant, negative impact on trust. Thus, although individual negative mood is less likely to fluctuate during a task, the general negative moods that people bring into a situation can have a significant effect on their level of trust in team members. Similar results for the impact of negative baseline mood on trust in automation have been found (Stokes and Lyons, 2009). However, research has shown that induced negative mood (i.e., change from baseline) also has a significant, negative impact on trust (Stokes et al., 2010).

Positive mood is believed to foster top-down processing of environmental stimuli (Bless et al., 1990); however, negative mood is believed to promote focused scanning of one's surroundings (Schwarz, 1990). These assertions and the findings of the current study relate to both mechanisms proposed in the AIM (Forgas, 1995). Individuals who experienced greater positive mood during the task may have used those positive moods as a heuristic shortcut to assume trust was warranted, thus supporting the faster cognitive processing proposed by the affect-as-information mechanism. Conversely, the negative mood of individuals likely prompted more in-depth processing, resulting in the affect-priming mechanism dominating. However, further research is needed to substantiate the separate activation of the AIM mechanisms based on positive and negative mood. Although additional research validating the mechanisms involved would add tremendous value to our understanding of the relationship between mood and trust, the findings in the present study confirm that individual mood, both positive and negative, significantly influences trust in multicultural team settings.

### *Team Mood Effects*

A similar pattern of results emerged when considering the link between team mood and trust. Positive mood at the team level was expected to be related to higher trust. This hypothesis was supported. Even after controlling for baseline and individual positive mood during the experiment, individuals' perceptions of team positive mood added significant variance when predicting trust. Similarly, negative team mood was expected to be related to lower trust. This hypothesis was also supported. Consistent with the notion that emotions signify evolutionary processes, triggering approach/avoidance tendencies (Lazarus, 1999), the current study demonstrated that individuals' reports of team mood influenced the degree to which individuals trusted their teammates. These findings support previous theoretical models which discuss emotions as an integral part of the trust process (McAllister, 1995; Schoorman et al., 2007; Williams, 2001, 2007), and, importantly, this relationship was found for both individual and team mood.

Previous studies have explored the relationship between individual mood and trust for both positive (Lount, 2010; Stokes et al., 2010) and negative mood (Stokes et al., 2010). Consistent with those studies, the current research found that both individual and team mood predicted trust. To our knowledge, this is the first study to examine the link between team mood and trust, notably in multicultural teams. One possible explanation for these findings comes from the AIM. The AIM posits that moods will infuse social and cognitive judgments in novel situations where the particular mood is not attributable to the situation (Forgas, 1995). The multicultural nature of the teams may have presented individuals with a novel scenario where they may have relied on their moods when making judgments about their team. A second explanation involves perceptions of trustworthiness. Affective processes may influence the degree to which individuals perceive others as trustworthy. For example, Williams (2001) suggests that moods cue the degree to which benevolence is perceived in others.

Several researchers suggest that moods influence the cognitive evaluations used to determine trustworthiness (Jones and George, 1998; Schoorman et al., 2007). This appears to be true of swift trust, the trust that initially forms during team interactions (Jarvenpaa and Leidner, 1999). However, these effects may be transient. Previous research which manipulated positive and negative mood found that individuals in a positively primed condition reported higher trust in automation during a decision making task, whereas those in a negatively primed condition reported lower trust of automation (Stokes et al., 2010). However, Stokes and colleagues (2010) found that the mood manipulation only shaped participants' initial trust perceptions. Webber (2008) suggests that trust evolves over time, with implications for both affective and cognitive mechanisms. Alternatively, Levin, Whitener, and Cross (2006) found that the antecedents of trust change over time, with dispositional factors driving trust perceptions early on and behavior (i.e., trustworthiness) impacting subsequent trust perceptions. In any event, mood appears to influence the degree of trust perceived by individuals during team-based tasks. Not only does an individual's current mood during a task influence trust of team members, but the perceived mood of the team as a whole also has a significant impact on team trust above and beyond that of individual mood.

Although much speculation exists regarding relationships between mood and trust, the findings of the present study add to the mounting empirical evidence (Dunn and Schweitzer, 2005; Lount, 2010; Stokes et al., 2010) supporting a more holistic view of the relationship. As opposed to the view that mood primarily functions as a factor in the deeper levels of affect-based trust (McAllister, 1995), our results clearly indicate that mood also impacts the early forms of cognitive-based trust. These findings are theoretically aligned with the AIM model (Forgas, 1995) for the impact of mood on judgment and decision making, thereby fusing trust research with the long history of social cognitive and psychological literature on risk-taking tendencies. These findings hold several implications for research and practice alike.

### *Implications*

Although there are many other factors that may help shape such early forms of trust (e.g., trust propensity, contextual cues), mood is one such factor that itself can be shaped, and therefore offers a leverage point for increasing trust in military teams. Military teams face the onerous challenges of modern warfare: high ops tempos (Huffman, Adler, Dolan, and Castro, 2005), novel stressors from peacekeeping and evolving roles (Bartone, Vaitkus, and Adler, 1998), distributed interactions which often impede trust formation (Ritter, Lyons, and Swindler, 2007), and team reconfiguration. The added difficulty of confronting many of these challenges in a multinational team setting (Klein and Steele-Johnson, 2007) makes team trust a critical component to team and mission success. Managing perceptions of team mood, particularly in newly developed teams, offers one mechanism for increasing trust in military teams. This trust in turn, increases team performance (Aubert and Kelsey, 2003; Dirks, 1999; Fuller et al., 2008; Jarvenpaa and Leidner, 1999; Jones and George, 1998; Lyons, Stokes, and Schneider, this volume).

### *Limitations and Future Research*

There are a few limitations in the current study. First, previous research suggests that moods, notably positive moods, tend to have a beneficial impact on trust perceptions only when indicators of trustworthiness are present (Lount, 2010). The current study did not explicitly provide or manipulate indicators of trustworthiness. Secondly, as stated above, trust evolves over time. The current study demonstrated that moods influence swift trust, or the trust that forms early in a team's lifecycle. While this is consistent with previous research (Stokes et al., 2010), it is unknown how moods influence trust throughout the latter interactions within a team. In fact, recent studies have shown that despite initial decrements in trust, virtual teams eventually establish trust levels that are commensurate with face-to-face teams when given enough time to establish team norms (Wilson, Straus, and McEvily, 2006). Third, the present study did not specifically measure cultural aspects outside of examining trust in multicultural teams. Future research should examine specific cultural variables to develop a deeper understanding of how trust operates in multicultural settings. For example, Lin (2010) found that differences in analytic and holistic thinking were related to trust in organizational scenarios. Analytic thinking is more common in individualistic or Western cultures, and holistic thinking is more common in collectivistic or Asian cultures (Lin and Klein, 2008). Finally, the current study utilized a highly interdependent team task. Researchers suggest that the relationship between trust and performance diminishes as interdependence is lessened (Dirks, 1999). It is unclear how the current results will generalize to less interdependent tasks. However, many military team tasks are highly interdependent. Future research should incorporate tasks of varying durations to explore the relationship of moods and trust at different time points within team interactions. Further, research should differentiate the relationship between moods and trust versus moods and trustworthiness.

*Conclusion*

Multicultural team settings have become a mainstay in military environments. Although there are many advantages to such teams, there are many challenges as well. Trust offers one possible mechanism for overcoming many of these challenges, as trust leads to improved team coordination and performance (Aubert and Kelsey, 2003; Dirks, 1999; Fuller, Mayer, and Pike, 2008; Jarvenpaa and Leidner, 1999; Jones and George, 1998; Lyons et al., this volume). However, enhancing trust in multicultural teams poses an added difficulty given the varied background of team members. Focusing on the cognitive underpinnings of trust and its formation offers a potential means of increasing trust across cultural barriers. The findings of the present study indicate that mood influences trust based on various cognitive information-processing mechanisms. Both individual mood and perceived team mood significantly contributed to the trust of team members from differing cultural backgrounds. Thus, managing perceptions of mood, particularly team mood, affords a relatively simple means of increasing trust, and in turn performance, in multicultural team settings.

## Acknowledgments

The present research was supported by the Air Force Research Laboratory contract, FA8650-04-D-6546, 0006. The authors would like to thank Captain Jason Seyba and David Ames for their programming support in developing the CAPS system. The authors would also like to thank Ms. Gaea Payton for her data collection support.

## References

Aubert, B.A. and Kelsey, B.L. (2003). Further understanding of trust and performance in virtual teams. *Small Group Research, 34*, 575–618.

Bartone, P.T., Vaitkus, M.A., and Adler, A.B. (1998). Dimensions of psychological stress in peacekeeping operations. *Military Medicine, 163*, 587–593.

Bless, H., Bohner, G., Schwarz, N., and Strack, F. (1990). Mood and persuasion: A cognitive response analysis. *Personality and Social Psychology Bulletin, 16*, 331–345.

Chou, K.L., Lee, T.M.C., and Ho, A.H.Y (2007). Does mood state change risk taking tendency in older adults? *Psychology and Aging, 22*(2), 310–318.

Colquitt, J.A., Scott, B.A., and LePine, J.A. (2007). Trust, trustworthiness, and trust propensity: A meta-analytic test of their unique relationships with risk taking and job performance. *Journal of Applied Psychology*, *92*, 900–927.

Cummings, L.L. and Bromiley, P. (1996). The organizational trust inventory (OTI): Development and validation. In R.M. Kramer and T.R. Tyler (eds), *Trust in Organizations: Frontiers of Theory and Research*, 302–330. Thousand Oaks, CA: Sage.

Dirks, K.T. (1999). The effects of interpersonal trust on work group performance. *Journal of Applied Psychology*, *84*, 445–455.

Dunn, J.R. and Schweitzer, M.E. 2005. Feeling and believing: The influence of emotion on trust. *Journal of Personality and Social Psychology*, *88*, 736–748.

Forgas, J.P. (1995). Mood and judgment: The affect infusion model (AIM). *Psychological Bulletin*, *117*, 39–66.

Forgas, J.P. and East, R. (2008). On being happy and gullible: Mood effects on skepticism and the detection of deception. *Journal of Experimental Social Psychology*, *44*, 1362–1367.

Fuller, M.A., Mayer, R.C., and Pike, R.E. (2008). Media and familiarity effects on assessing trustworthiness: 'What did they mean by that?' In L.L. Brennan and V.E. Johnson (eds), *Computer-mediated Relationships and Trust: Managerial and Organizational Effects*, 56–74. Hershey, NY: Information Science Reference.

Gaudry, E., Vagg, P., and Spielberger, C.D. (1975). Validation of the state-trait distinction in anxiety research. *Multivariate Behavioral Research*, *10*(3), 331–341.

George, J.M. (1990). Personality, affect, and behavior in groups. *Journal of Applied Psychology*, *75*, 107–116.

Hatfield, E., Cacioppo, J.T., and Rapson, R.L. (1993). Emotional contagion. *Current Directions in Psychological Science*, *2*, 96–99.

Huffman, A.H., Adler, A.B., Dolan, C.A., and Castro, C.A. (2005). The impact of operations tempo on turnover intentions of army personnel. *Military Psychology*, *17*(3), 175–202.

Ilies, R., Wagner, D.T., and Morgeson, F.P. (2007). Explaining affective linkages in teams: Individual differences in susceptibility to contagion and individualism-collectivism. *Journal of Applied Psychology*, *92*, 1140–1148.

Jarvenpaa, S.L. and Leidner, D.E. (1999). Communication and trust in global virtual teams. *Organizational Science*, *10*, 791–815.

Jones, G.R. and George, J.M. (1998). The experience and evolution of trust: Implications for cooperation and teamwork. *Academy of Management Review*, *23*(3), 531–546.

Kelly, J.R. and Barsade, S.G. (2001). Mood and emotions in small groups and work teams. *Organizational Behavior and Human Decision Processes*, *86*(1), 99–130.

Klein, H.A. and Steele-Johnson, D. (2007). Training for multinational teamwork. In R.R. Hoffman (ed.) *Expertise Out of Context: Proceedings of the Sixth International Conference on Naturalistic Decision Making*, 473–506. Mahwah, NJ: Lawrence Erlbaum.

Kramer, R.M. (1999). Trust and distrust in organizations: Emerging perspectives, enduring questions. *Annual Review of Psychology*, *50*, 569–589.

Lang, P.J., Bradley, M.M., and Cuthbert, B.N. (1990). Emotion, attention, and the startle reflex. *Psychological Review*, *97*, 377–395.

Lang, P.J., Bradley, M.M., and Cuthbert, B.N. (2008). International affective picture system (IAPS): Affective ratings of pictures and instruction manual. Technical Report A-8, University of Florida, Gainesville, FL.

Lazarus, R.S. (1999). *Stress and Emotions: A New Synthesis.* New York: Springer Publishing Company.

Lerner, J.S. and Keltner, D. (2000). Beyond valence: Toward a model of emotion-specific influences on judgment and choice. *Cognition and Emotion, 14(4)*, 473–493.

Levin, D.Z., Whitener, E.M., and Cross, R. (2006). Perceived trustworthiness of knowledge sources: The moderating impact of relationship length. *Journal of Applied Psychology*, *91*(5), 1163–1171.

Lin, M. (2010). Analytic-holistic thinking and personality factors affecting information processes during sensemaking. Poster presented at *The 118th American Psychological Association* (APA) Convention, San Diego, CA.

Lin, M. and Klein, H.A. (2008). Athletes, murderers, and a Chinese farmer: Cultural perspectives on sensemaking. In J.M.C Schraagen, L. Militello, T. Ormerod, and R. Lipshitz (eds), *Naturalistic Decision Making and Macrocognition*, 159–181. Aldershot, UK: Ashgate.

Lount, R.B., Jr. (2010). The impact of positive mood on trust in interpersonal and intergroup interactions. *Journal of Personality and Social Psychology*, *98*(3), 420–433.

Lyons, J.B., Stokes, C.K., Palumbo, M., Boyle, E., Seyba, J., and Ames, D. (2008). Bringing logistics into the laboratory: Developing a team-based logistics task. *Air Force Journal of Logistics*, *31*(4), 47–52.

Lyons, J.B., Stokes, C.K., and Schneider, T.R. (this volume). Predictors and outcomes of trust in teams. In N.A. Stanton (ed.), *Trust in Military Teams*.

Mayer, R.C., Davis, J.H., and Schoorman, F.D. (1995). An integration model of organizational trust. *Academy of Management Review*, *20*, 709–734.

McAllister, D.J. (1995). Affect- and cognition-based trust as foundations for interpersonal cooperation in organizations. *Academy of Management Journal*, *38*, 24–59.

Merritt, S.M. and Ilgen, D.R. (2008). Not all trust is created equal: Dispositional and history-based trust in human-automation interactions. *Human Factors*, *50*, 194–210.

Muir, B.M. (1994). Trust in automation: Part I. Theoretical issues in the study of trust and human intervention in automated systems. *Ergonomics*, *37*(11), 1905–1922.

Nauqin, C.E. and Paulson, G.D. (2003). Online bargaining and interpersonal trust. *Journal of Applied Psychology*, *88*, 113–120.

Nowlis, V., 1970. Mood: Behavior and experience. In M. Arnold (ed.), *Feelings and Emotions*, 261–277. New York: Academic Press.

Rempel, J.K., Holmes, J.G., and Zanna, M.P. (1985). Trust in close relationships. *Journal of Personality and Social Psychology*, *49*(1), 95–112.

Ritter, J., Lyons, J.B, Swindler, S.D. (2007). Large-scale coordination: Developing a framework to evaluate socio-technical and collaborative issues. *Cognition, Technology, and Work*, *9*(1), 33–38.

Schoorman, R.D., Mayer, R.C., and Davis, J.H. (2007). An integrative model of organizational trust: Past, present, and future. *Academy of Management Review*, *32*(2), 344–354.

Schwarz, N. (1990). Feelings as information: Informational and motivational functions of affective states. In E.T. Higgins and R. Sorrentino (eds), *Handbook of Motivation and Cognition: Foundations of Social Behavior*, *2*, 527–561. New York: Guilford.

Schwarz, N. and Clore, G.L. (1983). Mood, misattribution, and judgments of well-being: Informative and directive functions of affective states. *Journal of Personality and Social Psychology*, *45*, 513–523.

Schwarz, N., Bless, H., and Bohner, G. (1991). Mood and persuasion: Affective states influence the processing of persuasive communication. In M.P. Zanna (ed.), *Advances in experimental social psychology*, *24*, 161–199. San Diego, CA: Academic Press.

Stokes, C.K. and Lyons, J.B. (2009). The influence of affect on perceptions of trust. Presented at *The 21st Annual Convention of the Association for Psychological Science*, special theme program, San Francisco, CA.

Stokes, C.K., Lyons, J.G., Littlejohn, K., Natarian, J., Case, E., and Speranza, N. (2010). Accounting for the human in cyberspace: Effects of mood on trust in automation. Proceedings of *The IEEE 2010 International Symposium on Collaborative Technologies and Systems*, Chicago, IL.

Tanis, M. and Postmes, T. (2005). Short communication: A social identity approach to trust: Interpersonal perception, group membership and trusting behavior. *European Journal of Social Psychology*, *35*, 413–424.

Tomlinson, E.C. and Mayer, R.C. (2009). The role of causal attribution dimensions in trust repair. *Academy of Management Review*, *34*(1), 85–104.

Watson, D., Clark, L.A., and Tellegen, A. (1988). Development and validation of brief measures of Positive and Negative Affect: The PANAS scales. *Journal of Personality and Social Psychology*, *54*, 1063–1070.

Webber, S.S. (2008). Development of cognitive and affective trust in teams: A longitudinal study. *Small Group Research*, *39*(6), 746–769.

Williams, M. (2001). In whom we trust: Group membership as an affective context for trust development. *Academy of Management Review*, *26*(3), 377–396.

Williams, M. (2007). Disentangling concepts: The role of affect in trust development and cooperation. Presented at the Annual Academy of Management Meetings, Philadelphia, PA.

Wilson, J.M., Straus, S.G., and McEvily, B. (2006). All in due time: The development of trust in computer-mediated and face-to-face teams. *Organization Behavior and Human Decision Processes*, *99*, 16–33.

Zimmermann, P., Guttormsen, S., Danuser, B., and Gomez, P. (2003). Affective computing: A rationale for measuring mood with mouse and keyboard. *International Journal of Occupational Safety and Ergonomics*, *9*(4), 539–551.

# Chapter 3
# Predictors and Outcomes of Trust in Teams

Joseph B. Lyons and Charlene K. Stokes
*US Air Force Research Laboratory, Dayton, OH, USA*

Tamera R. Schneider
*Wright State University, Dayton, OH, USA*

## Background

Organizations in today's work environment are increasingly using teams to accomplish their goals and to compete in volatile markets (Cohen and Bailey, 1997; Kozlowski and Bell, 2003; Stevens and Campion, 1999). Due to increased globalization and advances in information technology (IT), many of these teams are characterized by geographic dispersion and thus are required to interact using computer-mediated mechanisms (Hertel, Geister, and Konradt, 2005). These virtual team arrangements can be positive because they allow organizations to span national and temporal boundaries, thus promoting flexibility and agility. However, distributed teams tend to suffer from a variety of negative consequences, such as reduced trust (Jarvenpaa and Leidner, 1999), impaired ability to resolve conflict (Gallupe, DeSanctis, and Dickson, 1988), and lower overall performance (Straus, 1997) when compared to face-to-face (FTF) teams. This presents challenges to employees working in distributed teams and to organizations that are forced to explore mechanisms to counteract the inextricable consequences of technology-mediated interactions. The military faces a similar problem, as soldiers often interact with others across the globe, oftentimes with little or no prior experience with the individuals on the other end. One potential solution to this issue involves identifying dispositions that are more or less conducive to facilitating trust in distributed team settings. To date, there is a dearth of research exploring the dispositional antecedents of trust in distributed teams while also considering relevant outcomes. The purpose of the current research was to examine dispositional antecedents (personality and cognitive ability) and outcomes (cohesion and performance) of trust in a computer-mediated team scenario.

Interpersonal trust has both cognitive and affective underpinnings (McAllister, 1995), but broadly represents the willingness of one person to make themselves vulnerable to the actions of another person (Mayer, Davis, and Schoorman, 1995). Trust, while beneficial to organizations and teams for a variety of reasons, is an elusive construct. To expound the trust construct, researchers have distinguished *trust* from *trustworthiness* and *trust propensity*, both of which are believed to be antecedents to accepting vulnerabilities to others' actions (i.e., trust). Trustworthiness is represented by variables such as ability, benevolence, and integrity, and typically individuals who possess higher levels of these characteristics are more likely to be trusted by others (Mayer et al., 1995). Indicators of trustworthiness are likely to be target- or context-dependent. For example, trust of certain individuals may be based on their ability to perform given tasks, their perceived integrity, or a shared past history of a positive nature. Similarly, certain situations (e.g., emergency response situations) may create social imbalances resulting in more or less trust of individuals based on their status or rank. Trust propensity, on the other hand, represents an individual's dispositional tendency to trust others (Colquitt , Scott, and LePine, 2007; Mayer et al., 1995). Past research has found that these dispositional trust elements have a stronger relationship with initial trust perceptions, while ongoing interactions have a greater influence on latter trust development (Merritt and Ilgen, 2008). In addition to being dominant during the early stages of interaction, the propensity to trust others may influence interpersonal interactions in a broad rather than a context-specific sense, and as such, may have implications for enhancing overall distributed team effectiveness. It is imperative that researchers understand how individuals form initial trust perceptions as these initial perceptions may influence subsequent perceptions in a consistent manner (Dirks and Ferrin, 2001).

An individual's intention to accept vulnerability (i.e., to trust) is based on the positive expectations of the actions of those trusted (Colquitt et al., 2007). These positive expectations are typically based on prior interactions that shed light on another's level of trustworthiness. Ferrin, Dirks, and Shah (2006) asserted that the central determinant of trust is the interpersonal behavior of the trusted during prior interactions. Intuitively, positive prior interactions should lead to expectations of positive future actions on the part of the trusted. However, distributed teams often lack prior interactions as well as the non-verbal cues that individuals may use to infer trustworthiness (Aubert and Kelsey, 2003), thereby making it difficult for team members to calibrate their trust of other teammates. In unfamiliar contexts, where prior interactions are lacking (e.g., a distributed team context), the dispositional aspects of trust (e.g., trust propensity) and perhaps other dispositions related to trust are likely to become the most relevant antecedents of trust (Bigley and Pearce, 1998; Rotter, 1980). Therefore, the present study focused on dispositional antecedents of trust. The context of the present research differs from the vast majority of the trust literature, which has focused on prior interactions or trustworthiness as a determinant of trust (Bigley and Pearce, 1998; Ferrin et al., 2006). There is limited research guidance in regard to dispositional predictors of trust (Merritt and Ilgen, 2008). However, the Industrial/Organizational Psychology literature often considers cognitive ability (Schmidt and Hunter, 1998) and personality (Barrick, Mount, and Judge, 2001) as relevant dispositional predictors of job performance. Given that research posits trust as a critical enabler of team success (Fuller, Mayer, and Pike, 2008), research is needed to explore if typical dispositional variables are related to trust. The present research investigated whether these traditional predictors of job performance are related to trust in teams.

### *Dispositional Predictors of Trust*

Cognitive ability represents intelligence and an individual's aptitude for solving problems (Drasgow, 2003). Historical (Spearman, 1927) and contemporary views of cognitive ability (Carroll, 1993) attest to the presence of a single 'g' factor which can be measured by maximum performance tests. Personality, in contrast, is multidimensional. Contemporary views of personality adhere to the Five-Factor Model (FFM) theory of personality (Goldberg, 1992; McCrae and Costa, 1987), where personality comprises five global dimensions: agreeableness, conscientiousness, extraversion, neuroticism, and openness. Research has found that cognitive ability (Schmidt and Hunter, 1998) and aspects of personality, most notably conscientiousness (Barrick et al., 2001), are robust predictors of job performance across a variety of settings. Additionally, cognitive ability and personality have received some attention in the team literature (Barrick, Stewart, Neubert, and Mount, 1998; Hollenbeck, Moon, Ellis, West, Ilgen, Sheppard, Porter, and Wagner, 2002; Williams and Sternberg, 1988). Barrick and colleagues (1998) found that teams with higher aggregates of cognitive ability, conscientiousness, extraversion, and emotional stability evidenced higher overall performance. Further, research by Williams and Sternberg (1988) found that heterogeneity in cognitive ability and homogeneity in personality predicted performance in team settings. However, the relationship between these variables and trust in team settings was not examined.

Bigley and Pearce (1998) reviewed the trust literature and found a common conceptualization of trust as a conscious decision to cooperate with other individuals. Individuals with high levels of cognitive ability may be more aware of the demands of teamwork and thus understand that trust and cooperation are essential components of teamwork. Recognition of these 'teamwork-related competencies' is not necessarily straightforward and may involve complex cognitive processing, suggesting that higher cognitive ability may be critical for recognizing and implementing good teamwork strategies. Another aspect of trust is a *cognitive* process that aids in understanding relationships among individuals (Lewis and Weigert, 1985). This aspect of trust may be grounded in cognitive judgments of another's competence or reliability (McAllister, 1995). Higher cognitive ability may enhance the cognitive calculations needed to determine the trustworthiness of others and enhance the recognition that trust is crucial to teamwork. This may be particularly true in distributed teams where trust perceptions are not based on prior interactions, and unlike face-to-face interactions, are more challenging to calibrate due to the limited or computer-mediated cues available (Fuller et al., 2008).

Although empirical research to support a link between cognitive ability and trust is lacking, research does support a link between personality and trust. Personality traits associated with positive social interactions serve to foster trust (Barrick et al., 1998). Agreeableness and extraversion are intrinsically interpersonal (McCrae and Costa, 1989), and emotional stability (i.e., low neuroticism) increases the likelihood of positive social interactions (Barrick et al., 1998). These factors should be related to trust, particularly in a team context where social interactions drive effectiveness. Past research has found that shyness, jealously, and suspicion are all negatively related to trust (Couch and Jones, 1997), thus demonstrating that dispositions can influence how individuals gauge trust perceptions.

The terms *trusting* and *cooperative* are nearly always included in the descriptive characteristics listed for agreeableness (McCrae and Costa, 1987). Agreeableness captures the relatively stable, general views and beliefs that individuals hold toward others in society, such as having positive expectations of others' actions, as well as characterizes one's general approach toward others. A meta-analysis found that agreeableness was the strongest FFM predictor of performance in interpersonal jobs, specifically those involving cooperative teamwork (Mount, Barrick, and Stewart, 1998). Trust propensity as a dispositional trait can be thought of as one aspect of agreeableness, focused specifically on trust in cooperative interactions. Given this similarity in conceptualization, agreeableness was expected to have a positive relationship with trust.

Past research has found a link between acceptance of vulnerability and extraversion. Nicholson, Soane, Fenton-O'Creevy, and Willman (2005) found that extraversion predicted risky behavior across six risk-taking domains: recreation, health, career, finance, safety, and social. In accord with the definition of trust as accepting vulnerability from others, past research has also found that extraverts are more likely to accept vulnerability in certain social situations (i.e., sexual encounters; Miller, Lyman, Zimmerman, Logan, Leukefeld, and Clayton, 2004). Specifically, those high in extraversion were more likely to make themselves vulnerable to the potential negative consequences associated with risky sexual behaviors. Past research has also found that extraversion is related to more trust in automated systems (Merritt and Ilgen, 2008). Because extraverts have a general tendency to accept vulnerability, we expected extraversion to be positively related to trust.

Neuroticism (low emotional stability) is negatively related to cooperation, an aspect of effective teamwork (Mount et al., 1998). Neuroticism is also associated with increased threat evaluations (Schneider, 2004) and, as such, may reduce the willingness to accept vulnerability in a given situation. The latter is necessary for trust to develop. Those high in neuroticism tend to engage in less pro-social behavior, which may result in suppressed cooperation and inhibited trust from others (Mount et al., 1998). High neuroticism may create a downward spiral of distrust in social situations. That is, initial inhibitive behavior and negative expectations reduce trusting behavior from others, which then provides substantive evidence for subsequent negative expectations for future positive interactions. This may be detrimental in a distributed team context where initial expectations and dispositions can drive trust perceptions. Trust has also been positively related to perceived similarity, most notably to demographic similarity (Levin, Whitener, and Cross, 2006). Neuroticism has been negatively related to perceived similarity in experimental studies (Moss, Garivaldis, and Toukhsati, 2007). Consequently, neuroticism is likely to have a negative relationship with trust.

Contrary to neuroticism, researchers have also found a positive relationship between openness and acceptance of risk (Gallagher, 2005; Kowert and Hermann, 1997; Nicholson et al., 2005). However, unlike extraverts and those high in agreeableness, individuals high in openness do not share the same social orientation (McCrae and Costa, 1989). Although individuals high in openness are more willing to accept risk than individuals low in openness, it is unclear how well this risk tendency will translate to interpersonal interactions in a team context. An individual's tendency to be open to diverse ideas, art, and unconventional values, as well as being intellectual, imaginative, and curious could easily be fulfilled in relative isolation. Further, research has found no support for openness as a predictor of team performance (Mount et al., 1998).

Unlike the above personality facets, conscientiousness does not have a clear conceptual link to trust. Although conscientiousness has received strong support as a predictor of individual job performance (Barrick et al., 2001), agreeableness is a stronger predictor of performance in interpersonal settings (Mount et al., 1998). Conscientiousness relates more to task-related aspects of jobs and less to interpersonal aspects of jobs (Barrick et al., 1998). Therefore, we did not expect a relationship between conscientiousness and trust.

### *Outcomes of Trust*

The study of trust has burgeoned in recent years largely due to the notion that trust is believed to influence behavior and relevant work outcomes (Dirks and Ferrin, 2001, 2002). Trust reduces the costs associated with monitoring personnel (Aubert and Kelsey, 2003), which could reduce team workload. Research has also shown that trust influences motivational processes such as goal setting. Dirks (1999) found that when individuals had higher trust they emphasized team, rather than individual goals. Finally, trust likely helps to foster effective team processes such as communication. Past research demonstrated that individuals contribute and share information more often when they report higher trust of their teammates (Kimmerle, Cress, and Hesse, 2007). Given previous findings that highlight beneficial outcomes associated with increased trust, the current study focused on examining the relationship between trust and two relevant outcomes: performance and team cohesion.

Meta-analytic studies have found that trust predicts better task performance, more organizational citizenship behaviors, more positive work attitudes, and less counterproductive work behaviors among individuals (Colquitt et al., 2007; Dirks and Ferrin, 2002), albeit the relationship between trust and work outcomes tends to be stronger for attitudinal versus performance-based variables. The current study expanded on this literature by exploring the link between trust and adaptive performance. Adaptive performance is distinct from traditional performance in that it represents an individual's ability to respond to novel demands (Pulakos, Arad, Donovan, and Plamondon, 2000). Trust may be most crucial for teams in situations when there are not clear ground rules for interactions (Dirks and

Ferrin, 2001), and particularly in those situations requiring adaptability. In such situations, individuals must truly put themselves in a position of vulnerability, as there may not be guidelines in place to protect the individuals from being exploited by others. Past research has substantiated this claim by showing that trust has a stronger relationship with information sharing in low structure situations (Staples and Webster, 2008). The current study explored the link between trust and performance both in a scenario where task duties were routine and one in which task routines were novel (i.e., requiring adaptability).

In addition to performance, another relevant outcome is team cohesion. Past research has found that trust in leadership is related to satisfaction with the leader and more positive perceptions of the social exchanges between leaders and subordinates (Dirks and Ferrin, 2002). While cohesion was not explicitly examined in that study, it is plausible that trust in one's teammates will be linked to more positive attitudes of the social exchanges between team members. Cohesion represents one's attraction and desire to remain within a team and has been linked to higher team performance (Mullen and Cooper, 1994), as well as motivation and commitment (Kozlowski and Bell, 2003). Past research has found that trust was indirectly related to intentions to remain within the team (Staples and Webster, 2008). Although trust is said to help team members maintain positive attitudes about each other in computer-mediated environments (Fuller et al., 2008), it is unclear whether trust will directly influence team cohesion in a computer-mediated scenario.

The purpose of the present research was to explore the link between dispositional factors such as personality and cognitive ability with trust while examining relevant outcomes of trust in a computer-mediated team scenario. The relevant outcomes focused on in the current study included task performance, adaptive task performance, and team cohesion. It was expected that agreeableness, extraversion, and openness would be positively related to trust. It was expected that neuroticism would be negatively related to trust. Further, trust was expected to be related to higher performance (both typical and adaptive) and cohesion. Given the lack of empirical backing for the link between cognitive ability and trust, no explicit hypotheses were posited.

## Method

### *Participants*

One hundred seventy-five undergraduate psychology students from a Midwestern university participated in this study in exchange for partial course credit. The average age was 20.28 ($SD = 4.41$) and 72 per cent of the sample were female. The majority of students were Caucasian (63 per cent), followed by African American (28 per cent), and Asian (5 per cent).

*Materials*

*Cohesion* Participants rated seven items using a seven-point Likert scale (1 = Strongly Disagree, 7 = Strongly Agree) to assess their attraction toward and liking of the group. An example item was 'The team members all get along well.' Reliability was acceptable (alpha = .85).

*Cognitive ability* The Wonderlic Personnel Test was used to assess general cognitive ability (Wonderlic, 1983). The Wonderlic is a 12-minute timed test of general verbal, math, and analytical abilities. Test-retest reliabilities for the Wonderlic generally range from .82 to .94 (Wonderlic, 1983).

*International Personality Item Pool* A 50-item measure assessing the Big Five personality factors (agreeableness, conscientiousness, extraversion, neuroticism, and openness) was used as an index of personality (http://ipip.ori.org/ipip/). For each dimension, participants rated the extent to which 10 statements described them on five-point scales (1 = Very Inaccurate, 5 = Very Accurate). Cronbach's alpha coefficients were acceptable for agreeableness (.76), conscientiousness (.84), extraversion (.84), neuroticism (.82), and openness (.76).

*Interpersonal trust* Trust was measured using a modified version of the Organizational Trust Inventory (OTI; Cummings and Bromiley, 1996). Modified versions of this scale have been used reliably in past research (see Naquin and Paulson, 2000), as in this research (Cronbach's alpha was .86). Using a seven-point scale (1 = Strongly Disagree, 7 = Strongly Agree) participants rated 11 items on the degree of trust they felt toward their teammates during the task. Example items include, 'I feel that other team members performed the task with me honestly,' and 'I think other team members took advantage of my problems.'

*Performance* Individual task performance scores were calculated for each person based on requisite duties, which are described in more detail below. These scores captured three critical elements: (a) whether the individual completed their assigned task, (b) whether duties were performed in the appropriate sequence in relation to teammates' duties, and (c) whether the required information was communicated to teammates. Individual task performance scores were manually calculated for each aircraft during the first experimental session and then for an adaptive performance episode in the second session. The scores were then standardized to allow for comparison across aircraft and adaptive events and were then averaged for each session. Two different raters coded event and communication data from the scenario to gauge performance. Cohen's Kappa averaged .92 for these ratings, thus indicating that the raters had high consistency.

*Task*

A team-based laboratory task, Computer-based Aerial Port Simulation (CAPS), was the computer-mediated team scenario. The CAPS software is a computer-generated, multi-player simulation program which mimics the logistics operations associated with an aerial port squadron (Lyons, Stokes, Palumbo, Boyle, Seyba, and Ames, 2008). The CAPS program includes five networked PCs that comprise the task environment that participants use to perform the task, and a sixth PC serving as the experimenter station for data upload and scenario manipulation. In the CAPS program, one team is composed of five interdependent functional stations: (a) passenger services, (b) fleet services, (c) cargo services, (d) ramp services, and (e) air terminal operations flight (ATOF). Passenger services must process, load, and unload all passengers. Fleet services must clean the aircraft and stock the aircraft with meals and other comfort items. Cargo services must process in-bound and out-bound cargo, which includes sequencing palletized cargo for pick-up by ramp services. Ramp services transports and unloads in-bound palletized cargo in the cargo bay and loads out-bound cargo on the aircraft. The ATOF monitors and directs all activities in the aerial port, essentially serving as the hub through which all information regarding aircrafts is received, processed, and dispatched to the other four functional stations.

The stations are interdependent; for example, fleet services should not clean the aircraft until passenger services has unloaded all passengers. Similarly, cargo services should not process in-bound cargo until ramp services transports and unloads the cargo. Thus, participants must coordinate and communicate their individual activities to achieve the shared goal of preparing aircraft for takeoff at or before its scheduled departure time. Steiner (1972) would label this type of task as conjunctive because it requires a minimum level of performance among all participants for successful task completion. Due to the high degree of communication required to complete this task, a vital component of the CAPS software is the instant message (IM) system. Participants are able to communicate needed information to other team members individually (to one other team member) or globally (to the entire team).

*Procedure*

Participants were run in teams of five. Upon arrival, an informed consent form was signed by each participant. Then participants were randomly assigned to one of the five computer stations. Each station was surrounded by partitions so that participants could not see any of the other participants. Participants were instructed that they were to use the IM chat system when communicating with their teammates, rather than talking directly. Participants also wore headphones to discourage overt verbal communication. Following their randomized role assignment, participants completed the Wonderlic Personnel Test. Participants then completed demographic and personality items that were presented on the computer monitor. Following these background questionnaires, participants were given a brief overview of the CAPS system and they began a training session. The training session consisted of three segments, general training, specific training, and a practice session. The general and specific training involved a self-paced, interactive PowerPoint presentation. Participants were given information about the CAPS system and aerial port operations in general, as well as more specific information relating to their unique role in the CAPS scenario. Following both the general and specific training segments, participants answered a few questions to gauge their engagement in the training material and their knowledge of the CAPS system. After the slide review, a practice trial commenced. During this time, the research assistant visited participant stations individually to answer participants' questions and, if necessary, to walk participants through the activities for their particular role. Following the training session, participants engaged in two 30-minute data collection sessions as part of a larger study. Trust was assessed after each 30-minute data collection session. The first session constituted the 'typical' performance period as the task duties remained constant. However, the second session presented participants with a novel scenario where they had to change the destination, cargo, and passengers of a current plane for a new destination with new cargo and passengers. To increase the adaptive requirements in session two, participants were faced with a communication breakdown, where previously used communication pathways were no longer functional and participants had to figure out the available pathways to convey the necessary information. Following both data collection sessions, participants were debriefed and released.

## Results

Table 3.1 shows the correlations between all the study variables, as well as descriptive statistics. As shown in Table 3.1, trust was related to cognitive ability, agreeableness, and openness, as expected. Unexpectedly, trust was not related to extraversion or neuroticism. Given the small to moderate interrelationships among the personality dimensions, a multiple regression analysis was used to identify the unique effects of the dispositional variables on trust. As shown in Table 3.2, cognitive ability and personality accounted for 15 per cent of the variance in trust perceptions, $F(6, 174) = 5.11$, $p < .001$. Cognitive ability, agreeableness, and openness each uniquely contributed to the prediction, and each factor predicted higher trust, as expected. In terms of outcomes measures, Table 3.1 shows that trust was related to higher performance for both routine task performance ($r = .20$) and adaptive task performance ($r = .24$), as expected. Also as expected, trust was related to higher team cohesion ($r = .54$).

**Table 3.1 Descriptive statistics and correlations among all study variables**

| Variable ($N$ = 175) | M | SD | 1 | 2 | 3 | 4 | 5 | 6 | 7 | 8 | 9 |
|---|---|---|---|---|---|---|---|---|---|---|---|
| 1. Agreeableness | 5.08 | .80 | *(.76)* | | | | | | | | |
| 2. Conscientiousness | 4.94 | .91 | .50** | *(.84)* | | | | | | | |
| 3. Extraversion | 4.84 | 1.00 | .12 | .30** | *(.84)* | | | | | | |
| 4. Neuroticism | 3.14 | .99 | -.43** | -.41** | -.36** | *(.82)* | | | | | |
| 5. Openness | 4.95 | .79 | .03 | .05 | .13 | .04 | (.76) | | | | |
| 6. Cognitive ability | 21.26 | 6.43 | .00 | .04 | -.01 | -.11 | .29** | | | | |
| 7. Trust | 5.40 | .94 | .26** | .10 | .00 | -.05 | .24** | .27** | *(.86)* | | |
| 8. Cohesion | 5.03 | 1.04 | .07 | .08 | .03 | .04 | .14 | .03 | .54** | *(.86)* | |
| 9. Task performance | .02 | .74 | .07 | .05 | .16* | -.08 | .13 | .37** | .20* | .15 | |
| 10. Adaptive performance | .06 | 1.02 | .09 | .04 | -.01 | -.03 | .07 | .29** | .24** | .09 | .51** |

*Note*: * $p < .05$. ** $p < .01$. Cronbach's Alpha shown on the diagonal (italicized).

**Table 3.2 Multiple regression results with cognitive ability and personality as predictors of trust**

| | $\beta$ | $R^2$ |
|---|---|---|
| Agreeableness | .25** | .15** |
| Conscientiousness | -.01 | |
| Extraversion | -.02 | |
| Neuroticism | .06 | |
| Openness | .16* | |
| Cognitive ability | .23** | |

*Note*: * $p < .05$. ** $p < .01$.

## Discussion

The present study was among the first to examine the concurrent influence of cognitive ability and personality on trust perceptions in a distributed team context. It was expected that agreeableness, extraversion, neuroticism, and openness would be positively related to trust. These expectations were partially supported. Agreeableness and openness were both found to be unique predictors of trust. Past research has shown that agreeableness is predictive of team performance (Barrick et al., 1998; Mount et al., 1998). The present study extends previous research by showing that agreeableness is the strongest FFM predictor of trust in a computer-mediated team setting. This is consistent with theoretical views of agreeableness which posit linkages to cooperation and trust (McCrae and Costa, 1987). Our research suggests that agreeableness is a practical variable to consider when devising teams for computer-mediated teamwork. Openness was also related to trust. This is consistent with past research that has established a link between openness and risk taking (Gallagher, 2005; Kowert and Hermann, 1997; Nicholson et al., 2005). The current study has extended previous research by demonstrating that openness generalizes to acceptance of risk in social situations.

Surprisingly, neither extraversion nor neuroticism were related to trust in the current study. This is inconsistent with past research which has shown a link between extraversion and risk taking (Miller et al., 2004; Nicholson et al., 2005). Research has found that extraverts tend more toward risks that entail physical harm rather than economic loss (Kowert and Hermann, 1997), suggesting that the influence of extraversion on trust may be moderated by the type of activity being performed. The virtual nature of the current study – with individuals engaged in teamwork using a computer-mediated scenario rather than interacting face-to-face – may have masked any potential relationship between extraversion and trust. This virtual interaction may have attenuated any relationship of extraversion and trust perceptions due to a lack of non-verbal cues. The lack of a relationship between trust and neuroticism was surprising. Past research has linked neuroticism with increased appraisals of threat (Schneider, 2004), which intuitively would make an individual less inclined to accept further vulnerabilities from others. A possible explanation is that the computer-mediated scenario may have prevented highly neurotic individuals from engaging in neurotic behavior. Past research has shown that when teams interact using computer-mediated means, they are less likely to exhibit negative socio-emotional communication (Nam, Lyons, Hwang, and Kim, 2009), thus it may have been that neurotic individuals' overt negativity may have been masked by the computer-mediated scenario. Future research is needed to clarify the relationship between these aspects of personality and trust.

Interestingly, individuals with higher cognitive ability reported higher trust perceptions of their teammates during the computer-mediated task. This finding remained when considering the impact of personality on trust. Past research has shown that teams with higher aggregated cognitive ability scores outperform those with lower scores (Barrick et al., 1998; Neuman and Wright, 1999). It is possible that the heightened trust fosters better performance in teams with higher cognitive ability. However, it is also possible that being a member of a high performance team results in higher trust.

To our knowledge, research has yet to examine how broad dispositional variables such as cognitive ability and personality influence trust perceptions in a computer-mediated scenario. Given that distributed teams may at times lack prior face-to-face interactions, dispositional predictors of trust become paramount. Past research has typically examined trust development based on prior interactions (Ferrin et al., 2006). In contrast, the present study examined dispositional variables of the trustee to explore how these factors influenced one's tendency to trust others in a computer-mediated scenario. The results of the present study clearly identify cognitive ability, agreeableness, and openness as dispositional correlates of trust.

Cognitive ability, with agreeableness and openness, may serve to capture, in part, the cognitive and non-cognitive components of trust (McAllister, 1995). Determining the trustworthiness of others in such situations is likely to be complex, requiring greater cognitive ability because of the lack of prior historical information (Lewis and Weigert, 1985). However, beyond the capacity of discerning trustworthiness in others, one may also need an inherent propensity to get along with others in order to establish trust among distributed team members, and an openness to experience new types of interactions. While speculative, these results offer support for the theory of trust as involving both cognitive and non-cognitive processes (McAllister, 1995).

Trust was expected to be related to several outcome measures, including performance and cohesion. These hypotheses were supported. Trust was related to both typical and adaptive performance, as expected. This is consistent with a previous meta-analysis which found a positive relationship between trust and task performance (Colquitt et al., 2007). The current study extends previous research by showing that trust is related to performance in a scenario requiring team adaptability. Adaptability is an emerging theme in the teams literature (Burke, Stagl, Salas, Pierce, and Kendall, 2006). Burke and colleagues discuss five enablers of adaptability in teams: performance monitoring, back-up behaviors, communication, leadership, and coordination. Trust may facilitate team adaptability by reducing the need for extensive performance monitoring (Aubert and Kelsey, 2003), creating the psychological safety for risk taking and information sharing (Kimmerle et al., 2007), and facilitating a team perspective (Dirks, 1999). Trust was also related to team cohesion, as expected. Trust in team settings fosters a collective, rather than an individual perspective (Dirks, 1999). In addition to these motivational benefits, the current study demonstrates that trust impacts team process. The current findings provide compelling evidence for the benefits of trust and they add to the growing literature on trust in team settings.

### *Limitations*

One limitation of the present study is that it is cross-sectional, rather than longitudinal. This limitation inhibits the exploration of these effects over time. Past research has found that during the early stages of team interactions, individuals tend to rely on demographic similarity to form trust perceptions, whereas trust is based more on trustworthy behaviors during the latter stages of team interactions (Levin et al., 2006). A second limitation of the present study is that individual contributions to the group task were not emphasized. The only visible performance metric, from the participant's perspective, was the time to process each aircraft (a team-level phenomenon). Past research has found that low sensation seekers (one facet of extraversion) were less likely to trust other team members when no group feedback was given about their individual performance, whereas high sensation seekers were more likely to trust when their individual contributions were made visible to the group (Kimmerle et al., 2007). Future research should explore how cognitive ability and the FFM dimensions relate to trust under varying degrees of feedback. Additionally, future research should continue to explore dispositional factors outside of the traditional FFM of personality. For example, past research has found that belief in a just world is related to higher trust perceptions (Begue, 2002). Personal beliefs and other variables related to trust need to be considered in future research to better understand how our dispositions influence trust perceptions.

## Conclusion

The present study examined the effects of dispositional variables on trust perceptions in a distributed team scenario. Rapid globalization and increased use of virtual teams represent catalysts for researchers to explore how certain dispositions relate to trust perceptions. Computer-mediated interactions have become a ubiquitous aspect of modern work life and military operations, yet little is known about how to overcome their inherent limitations. The present research revealed three dispositional predictors of trust that can offer guidance for future research and team selection/training. While this is a promising step in the exploration of how dispositions influence the trust process, further research is required to determine how trust can be maximized in virtual team settings. Further, the current study extended previous research by illuminating the trust–performance and trust–cohesion linkages. Trust was found to relate to both routine task performance and non-routine task performance, and this has strong implications for research and application.

## Acknowledgments

The present research was supported by the Air Force Research Laboratory contract, FA8650-04-D-6546, 0006. The authors would like to thank Captain Jason Seyba and David Ames for their programming support in developing the CAPS system. The authors would also like to thank Ms. Gaea Payton for her data collection support.

## References

Barrick, M.R., Mount, M.K., and Judge, T.A. (2001). Personality and performance at the beginning of the new millennium: What do we know and where do we go? *International Journal of Selection and Assessment*, *9*, 9–30.

Barrick, M.R., Stewart, G.L., Neubert, M.J., and Mount, M.K. (1998). Relating member ability and personality to work-team processes and team effectiveness. *Journal of Applied Psychology*, *83*, 377–391.

Begue, L. (2002). Beliefs in justice and faith in people: Just world, religiosity, and interpersonal trust. *Personality and Individual Differences*, *32*, 375–382.

Bigley, G.A. and Pearce, J.L. (1998). Straining for shared meaning in organizations science: Problems of trust and distrust. *Academy of Management Review*, *23*, 405–421.

Burke, C.S., Stagl, K.C., Salas, E., Pierce, L., and Kendall, D. (2006). Understanding team adaptation: A conceptual analysis and model. *Journal of Applied Psychology*, *91*, 1189–1207.

Carroll, J.B. (1993). *Human Cognitive Abilities: A Survey of Factor-Analytic Studies*, Cambridge, UK: Cambridge University Press.

Cohen, S.G. and Bailey, D.E. (1997). What makes teams work: Group effectiveness research from the shop floor to the executive suite. *Journal of Management*, *23*, 239–290.

Aubert, B.A. and Kelsey, B.L. (2003). Further understanding of trust and performance in virtual teams. *Small Group Research*, *34*, 575–618.

Colquitt, J.A., Scott, B.A., and LePine, J.A. (2007). Trust, trustworthiness, and trust propensity: A meta-analytic test of their unique relationships with risk taking and job performance. *Journal of Applied Psychology*, *92*, 900–927.

Couch, L.L. and Jones, W.H. (1997). Measuring levels of trust. *Journal of Research in Personality*, *31*, 319–366.

Cummings, L.L. and Bromiley, P. (1996). The organizational trust inventory (OTI): Development and validation. In R.M. Kramer and T.R. Tyler (eds), *Trust in Organizations: Frontiers of Theory and Research*, 302–330. Thousand Oaks, CA: Sage.

Dirks, K.T. (1999). The effects of interpersonal trust on work group performance. *Journal of Applied Psychology*, *84*, 445–455.

Dirks, K.T. and Ferrin, D.L. (2001). The role of trust in organizational settings. *Organizational Science*, *12*, 450–467.

Dirks, K.T. and Ferrin, D.L. (2002). Trust in leadership: Meta-analytic findings and implications for research and practice. *Journal of Applied Psychology*, *87*, 611–628.

Drasgow, F. (2003). Intelligence and the workplace. In W.C. Borman, D. R. Ilgen, and R. J. Klimoski (eds), *Handbook of Psychology: Industrial and Organizational Psychology*, *12*, 107–130. New York: John Wiley and Sons Inc.

Ferrin, D.L., Dirks, K.T., and Shah, P.P. (2006). Direct and indirect effects on third-party relationships on interpersonal trust. *Journal of Applied Psychology*, *91*, 870–883.

Fuller, M.A., Mayer, R.C., and Pike, R.E. (2008). Media and familiarity effects on assessing trustworthiness: 'What did they mean by that?' In L.L. Brennan and V.E. Johnson (eds), *Computer-mediated Relationships and Trust: Managerial and Organizational Effects*, 56–74, Hershey, NY: Information Science Reference.

Gallagher, M.E. (2005, November). High rolling leaders: The 'Big Five' Model of personality and risk-taking during war. http://www6.miami.edu/maia/ISAS05/papers/Gallagher_ISAS05.pdf, accessed October 31, 2007.

Gallupe, R.B., DeSanctis, G., and Dickson, G.W. (1988). Computer-based support for group problem finding: An experimental investigation. *MIS Quarterly*, *12*, 277–298.

Goldberg, L.R. (1992). The development of markers for the Big Five factor structure. *Psychological Assessment*, *4*, 26–42.

Hertel, G., Geister, S., and Konradt, U. (2005). Managing virtual teams: A review of the current empirical research. *Human Resource Management Review*, *15*, 69–95.

Hollenbeck, J.R., Moon, H., Ellis, A.P.J., West, B.J., Ilgen, D.R., Sheppard, L., Porter, C.O.L.H., and Wagner, J.A. (2002). Structural contingency theory and individual differences: Examination of external and internal person-team fit. *Journal of Applied Psychology*, *87*, 599–606.

International Personality Item Pool: A scientific collaboratory for the development of advanced measures of personality traits and other individual differences. http://ipip.ori.org/, accessed 21 June, 2005.

Jarvenpaa, S.L. and Leidner, D.E. (1999). Communication and trust in global virtual teams. *Organizational Science*, *10*, 791–815.

Kimmerle, J., Cress, U., and Hesse, F.W. (2007). An interactional perspective on group awareness: Alleviating the information-exchange dilemma (for everybody?). *International Journal of Human-Computer Studies*, *65*, 899–910.

Kowert, P.A. and Hermann, M.G. (1997). Who takes risks? Daring and caution in foreign policy making. *Journal of Conflict Resolution*, *41*, 611–637.

Kozlowski, S.W.J. and Bell, B.S. (2003). Work groups and teams in organizations. In W. Borman and D. Ilgen (eds), *Handbook of Psychology: Industrial and Organizational Psychology*, *12*, 333–375. New York: John Wiley and Sons Inc.

Levin, D.Z., Whitener, E.M., and Cross, R. (2006). Perceived trustworthiness of knowledge sources: The moderating impact of relationship length. *Journal of Applied Psychology*, *91*, 1163–1171.

Lewis, J.D. and Weigert, A. (1985). Trust as a social reality. *Social Forces*, *63*, 967–985.

Lyons, J.B., Stokes, C.K., Palumbo, M., Boyle, E., Seyba, J., and Ames, D. (2008). Bringing logistics into the laboratory: The development of a team-based logistics task. *Air Force Journal of Logistics*, *31*, 49–55.

Mayer, R.C., Davis, J.H., and Schoorman, F.D. (1995). An integration model of organizational trust. *Academy of Management Review*, *20*, 709–734.

McCrae, R.R. and Costa, Jr., P.T. (1987). Validation of the five-factor model of personality across instruments and observers. *Journal of Personality and Social Psychology*, *52*, 81–90.

McCrae, R.R. and Costa, Jr., P.T. (1989). The structure of interpersonal traits: Wiggins's circumplex and the Five-Factor Model. *Journal of Personality and Social Psychology*, *56*, 586–595.

McCrae, R.R. and John, O.P. (1992). An introduction to the Five-Factor Model and its applications. *Journal of Personality*, *60*, 175–215.

McAllister, D.J. (1995). Affect- and cognition-based trust as foundations for interpersonal cooperation in organizations. *Academy of Management Journal*, *38*, 24–59.

Merritt, S.M. and Ilgen, D.R. (2008). Not all trust is created equal: Dispositional and history-based trust in human-automation interactions. *Human Factors*, *50*, 194–210.

Miller, J.D., Lyman, D., Zimmerman, R.S., Logan, T.K., Leukefeld, C., and Clayton, R. (2004). The utility of the Five-Factor Model in understanding risky sexual behavior. *Personality and Individual Differences*, *36*, 1611–1626.

Moss, S.A., Garivaldis, F.J., and Toukhsati, S.R. (2007). The perceived similarity of other individuals: The contaminating effects of familiarity and neuroticism. *Personality and Individual Differences*, *43*, 401–412.

Mount, M.K., Barrick, M.R., and Stewart, G.L. (1998). Five-Factor Model of personality and performance in jobs involving interpersonal interactions. *Human Performance*, *11*, 145–167.

Mullen, B. and Cooper, C. (1994). The relation between group cohesiveness and performance: An integration. *Psychological Bulletin*, *115*, 210–227.

Nam, C.S., Lyons, J.B., Hwang, H., and Kim, S. (2009). The process of team communication in multi-cultural contexts: An empirical study using Bales' interaction process analysis (IPA). *International Journal of Industrial Ergonomics*, *39*, 771–782.

Nauqin, C.E. and Paulson, G.D. (2003). Online bargaining and interpersonal trust. *Journal of Applied Psychology, 88*, 113–120.

Neuman, G.A. and Wright, J. (1999). Team effectiveness: Beyond skills and cognitive ability. *Journal of Applied Psychology, 84*, 376–389.

Nicholson, N., Soane, E., Fenton-O'Creevy, M., and William, P. (2005). Personality and domain-specific risk taking. *Journal of Risk Research, 8*, 157–176.

Pulakos, E. D., Arad, S., Donovan, M. A., and Plamondon, K. E. (2000). Adaptability in the workplace: Development of a taxonomy of adaptive performance. *Journal of Applied Psychology, 85*, 612–624.

Rotter, J.B. (1980). Interpersonal trust, trustworthiness, and gullibility. *American Psychologist, 35*, 1–7.

Schmidt, F.L. and Hunter, J.E. (1998). The validity and utility of selection methods in personnel psychology: Practical and theoretical implications of 85 years of research findings. *Psychological Bulletin, 124*, 262–274.

Schneider, T.R. (2004). The role of neuroticism on psychological and physiological stress responses. *Journal of Experimental Social Psychology, 40*, 795–804.

Spearman, C. (1927). *The Abilities of Man*. New York: MacMillan.

Staples, S.D. and Webster, J. (2008). Exploring the effects of trust, task interdependence and virtualness on knowledge sharing in teams. *Information Systems, 18*, 617–640.

Steiner, I.D. (1972). *Group Processes and Productivity*. New York: Academic Press.

Stevens, M.J. and Campion, M.A. (1999). Staffing work teams: Development and validation of a selection test for teamwork settings. *Journal of Management, 25*, 207–228.

Straus, S.G. (1997). Technology, group process, and group outcomes: Testing the connections in computer-mediated and face-to-face groups. *Human-Computer Interaction, 12*, 227–266.

Williams, W.M. and Sternberg, R.J. (1988). Group intelligence: Why some groups are better than others. *Intelligence, 12*, 351–377.

Wonderlic, C. (1983). *Wonderlic Personnel Test Manual*, Northfield, IL: E.F. Wonderlic and Associates.

# Chapter 4
# Exploring the Impact of Cross-Training on Team Process

Joseph B. Lyons, Gregory J. Funke, Alex Nelson
and Benjamin A. Knott
*US Air Force Research Laboratory, Dayton, OH, USA*

### Background

Issues regarding teamwork have been a topic of interest within the research community for the last several decades. Teams comprise an essential capability within the Department of Defense (DoD) and may range from small combat teams to large and complex command and control (C2) teams. Despite their prominence within the DoD echelons, there is still a great deal to learn about team effectiveness. Breakdowns of team process can be catastrophic for military operations given the high tempo and dangerous mission environments that the DoD operates within. 'Military taskings can have critical time constraints, can involve high stakes, and must often be accomplished under conditions of mental fatigue, high workload, threat of physical injury, and in hostile environments' (McCann, Baranski, Thompson, and Pigeau, 2000, p. 1095). High profile accidents have occasionally been attributed to breakdowns in team processes, such as the destruction of Iran Air Flight 655 by the USS Vincennes (Johnston, Cannon-Bowers, and Salas, 1998). In that case, a military C2 team operating under high stress mistakenly identified a passenger plane as a hostile aircraft and engaged it with lethal force, resulting in nearly 300 civilian casualties. Incidents such as this illustrate that generating effective and efficient teamwork is imperative for the DoD. Understanding and predicting team effectiveness has also been an interest of academic researchers. This chapter will discuss team cross-training as a mechanism to foster effective team processes.

Most team taxonomies involve an Input–Process–Output (IPO) paradigm which generates considerations for team inputs, team processes, and team outcomes (Kozlowski and Bell, 2003). Team inputs can include variables that range from the personalities of team members to team composition factors (e.g., team size, diversity, etc.). Team process variables often revolve around issues of communication, cohesion, and trust. Finally, a typical team outcome variable of interest to researchers is team performance. Considering the IPO paradigm, the current study sought to explore the impact of team training (one example of a team input factor) on team process variables (i.e., communication, cohesion, and trust), on one motivational construct (collective efficacy), as well as on team performance.

### *Team Cross-Training*

Broadly defined, training represents a set of planned activities or programs designed by an organization to change employees' skills, attitudes, or behaviors (Cascio, 1998). A recent review of the team literature acknowledged the beneficial impact of team training on performance (Salas, Cooke, and Rosen, 2008). Much of this team training involves the development of teamwork skills such as conflict resolution and communication, thus suggesting that team process factors are critical for team effectiveness. A recent meta-analysis also supports the assertion that effective team training can result in enhanced team performance (Salas, Nichols, and Driskell, 2007). One such team training strategy involves cross-training. Cross-training represents an activity wherein individual team members are trained on the tasks, duties, and responsibilities of all other team members (Volpe, Cannon-Bowers, and Salas, 1996). Research suggests that teams that engage in cross-training evidence better communication and team performance relative to teams without such training (Cannon-Bowers, Salas, Blickensderfer, and Bowers, 1998; Volpe et al., 1996). Thus, cross-training may foster the development of effective team processes.

There are a variety of cross-training techniques. Blickensderfer and colleagues (1998) discussed three types of cross-training: positional clarification, modeling, and rotation. Positional clarification involves the verbal or written presentation of the roles and activities of the team. Positional modeling involves both the presentation of role information as well as the observation of the activities through either live modeling or video presentation. Finally, positional rotation occurs when team members actually participate in the different roles within the team. These three cross-training techniques differ in the complexity of the tactic, the level of involvement for the trainees, and the realism within which the trainees practice. Positional clarification involves the lowest complexity, least amount of trainee involvement, and lowest psychological fidelity relative to the other two methods. In contrast, positional rotation is the most complex, involves the highest degree of trainee involvement, and provides the trainees with the most realistic scenarios to engage in. The current research will focus primarily on positional rotation.

Cross-training may help team members anticipate the information needs of their teammates (McCann et al., 2000; Volpe et al., 1996), thereby promoting trust and cohesion between team members as they establish common ground and shared understanding of the task and context. Further, research has shown that cross-training results in the development of shared mental models, and that the relationship between shared mental models and team performance is mediated by coordination (Marks, Sabella, Burke, and Zaccaro, 2002). In other words, cross-training helps teams to better understand the environment, and this is primarily driven by their enhanced ability to coordinate with one another. Thus, team cross-training has implications for influencing team processes.

### *Team Processes*

The interactions (i.e., processes) that take place within a team are likely drivers of the success or failure of a team (Druskat and Kayes, 1999). Within the team and work group literature, perhaps the most discussed team process variable is communication. Communication represents the primary mechanism through which team members interact, and team training may be used to influence the development of team communication and interaction. As the complexities of modern warfare create heightened demands for military teams, information exchange between military operators remains a cornerstone of current and future military teams (Alberts and Hayes, 2003). Research has shown that high-performing teams tend to optimize information exchanges (Aubert and Kelsey, 2003). Taxonomies of team adaptability emphasize that communication is critical to team performance and flexibility (Burke, Stagl, Salas, Pierce, and Kendall, 2006). As an integral aspect of coordination, cross-training may also enhance team communication. Previous research suggests that cross-training results in enhanced coordination (Marks et al., 2002); as such, it was expected in this study that cross-training would have a similar advantageous effect on team communication.

Another relevant team process variable is cohesion, which can be thought of as the attraction of team members to the group. Cohesion has been linked to lower team conflict and higher team satisfaction (Kozlowski and Bell, 2003). Meta-analytic procedures have evidenced a small but positive relationship between cohesion and performance (Mullen and Cooper, 1994). Cohesion is socially oriented, thus the socialization of team members with one another may impact the degree to which the team develops cohesion. Team training techniques may reduce team conflict by establishing team norms, performance expectations, and feedback mechanisms. Cross-training may enhance team cohesion because of the shared understanding that develops within the team during cross-training activities. This shared understanding may help the team to establish a common ground for handling conflict, giving feedback, and managing performance expectations. These factors should, in turn, promote higher cohesion.

The willingness to share information and the ability to convince others to provide information are key elements that may differentiate high and low performing teams, and these elements are dependent on the construct of trust. Generally, trust represents the willingness of an individual to accept vulnerability to the actions of others (Mayer, Davis, and Schoorman, 1995). There are three fundamental aspects of trust. First, trust is relational. Trust is only relevant in the context of a relationship between two or more entities. These relationships may constitute interpersonal relations (Mayer et al., 1995) or interactions between a person and an information technology system (i.e., trust in automation; e.g., Lee and See, 2004; Parasuraman and Riley, 1997). Given this first assumption, trust is highly relevant in a team context. Second, research has shown that trust is dynamic and may change over the course of an interaction, with dispositional factors influencing trust early and indicators of trustworthiness influencing trust later in the process (Levin, Whitener, and Cross, 2006). This latter point is critical for team training given that different training techniques may foster different levels of exposure to team members' trustworthiness. Finally, trust is only relevant in the context of risk. If there is no risk in a situation then trusting others is irrelevant.

Past research suggests that trusting teammates may reduce the costs associated with monitoring personnel (Aubert and Kelsey, 2003), promoting collective rather than individual motivational tendencies (Dirks, 1999) and facilitating information sharing in team contexts (Kimmerle, Cress, and Hesse, 2007). Thus, trust is a highly relevant team process within military operational teams. Recent meta-analytic data supports positive relationships between trust and task performance, and trust and helping behavior (Colquitt, Scott, and LePine, 2007), both of which are relevant within military teams. Indicators of trustworthiness in interpersonal relationships include *ability*, *benevolence*, and *integrity* (Mayer et al., 1995). Accordingly, individuals may be trusted more when they possess high ability, are perceived as having benevolent intentions, and when they are thought to have high integrity. As perceptions of trustworthiness become the driving force for trust perceptions over time (Levin et al., 2006), it is plausible that the types of training a team engages in will impact trust development. Different training methods may provide team members with disparate information regarding the trustworthiness of other team members. Specifically, team cross-training techniques may provide team members with more diverse perspectives from which to gauge the ability, benevolence, and integrity of their teammates as they adopt and adapt to differing task and team roles.

### *Collective Efficacy*

Self-efficacy represents the belief that one possesses the ability to meet the demands of a specific situation (Bandura, 1997). These perceptions may be self-focused (evaluation of personal abilities), or they may be team-focused, as individuals evaluate the ability of their team to accomplish some goal. The latter is referred to as collective efficacy. Efficacy researchers (e.g., Walumbwa, Wang, Lawler, and Shi, 2004) have demonstrated that collective efficacy can be shaped by leadership, and is thus malleable. Additionally, research has shown that when individuals are learning-oriented they report higher efficacy during task situations (VandeWalle, Cron, and Slocum, 2001), which suggests that training may influence collective efficacy. Cross-training is an example of a learning-oriented activity, and as such may promote the development of collective efficacy. Similar to trust, team cross-training may foster collective efficacy by affording individuals more opportunities to calibrate their perceptions of team abilities from a variety of perspectives.

### *Team Performance*

As noted previously, researchers investigating the effects of cross-training on teams have found that such training may improve aspects of team process and task performance (e.g., Cannon-Bowers et al., 1998; Volpe et al., 1996), perhaps suggesting that the effects of cross-training on team performance are moderated by team process factors. By providing opportunities to foster the development of effective team processes, cross-training may afford teams more effective 'tools' with which to approach task performance, particularly in challenging situations (Cannon-Bowers et al., 1998). While some team processes may also influence each other (e.g., communication moderates trust, etc.), at a macroscopic level these processes necessarily support effective team performance. By providing novel perspectives and insights into the roles and responsibilities of teammates at an early stage of team development, cross-training may positively impact the subsequent development and maturation of team processes, resulting in improved team performance.

## Experimental Hypotheses

The overall objective of the present study was to explore the impact of cross-training on team processes in a simulated military command and control task over time. It was expected that cross-trained teams would report more effective communication, higher cohesion, greater trust of their teammates, and higher collective efficacy relative to individuals on teams that did not receive cross-training. In addition, it was hypothesized that cross-trained teams would perform a C2 task more effectively than non-cross-trained teams.

## Method

### *Participants*

Thirty participants completed a team-based experiment in five-person teams producing a total of six teams as part of a larger study. The composition of participants was 71 percent men and 29 percent women. Participants' ages ranged from 18 to 30 years ($M = 21.94$, $SD = 3.16$). Potential participants were screened using a biographical survey and the letter-sets test (Ekstrom, French, Harman, and Derman, 1976). Those who were selected for inclusion in the study were monetarily compensated for their participation.

### *Design*

Teams were randomly assigned to one of two treatment conditions (permanent or rotation) and participants were randomly given an initial assignment to one of three roles (weapons director, strike operator, tanker operator) within each team. Members of teams receiving the permanent treatment condition maintained constant roles throughout the entire experiment. In contrast, members of the rotation team condition were assigned new roles within their team for each new data collection session, and thereby received cross-training in each team role. During the experiment, teams completed five experimental sessions (one per day), enabling the further development of team processes and performance strategies.

### *Materials*

*Apparatus* This experiment employed Aptima, Inc.'s Distributed Dynamic Decision-making (DDD) software (version 3.0; MacMillan, Entin, Hess, and Paley, 2004). DDD provides a scriptable, low-to-moderate fidelity, team-in-the-loop simulated environment. DDD has successfully been used to simulate team command and control tasks and to study realistic and complex team processes in a variety of military and civilian research projects (MacMillan et al., 2004). The simulation required nine networked PCs communicating under TCP/IP protocol. Each workstation was equipped with a keyboard, mouse, and monitor. The DDD tactical display was projected to participants on a 17-inch Dell 1707FPv LCD monitor. Verbal communication, participant behavior, and screen captures were recorded using Nuance SDK (version 8.5) and MORAE (version 1.1.1) software.

In the experiment, five-person teams worked together to complete a simulated C2 air defense task. This task has been used in several previous experiments in military settings and has been demonstrated to be sensitive to experimental manipulations (e.g., Finomore, Knott, Nelson, Galster, and Bolia, 2007). Participants were randomly assigned to one of three team positions; positions differed in their roles and capabilities. The scenario required two weapons directors (WDs), two strike operators (SOs), and one tanker operator (TO). Within the simulation, the WDs' roles were to match friendly fighters with appropriate enemy targets, schedule fighters for refueling and resupply, and communicate their plans with other team members. The role of the strike and tanker operators was to maneuver team assets as instructed, to engage enemy targets, and to provide pertinent information to teammates concerning asset resources.

The air defense task was presented on a tactical display which included information about friendly and enemy assets. The simulation comprised three battle management zones: a 'preferred attack zone' where enemy assets entered the scenario, a sensitive 'friendly territory,' and a highly sensitive 'home territory' that contained an air base and ranger units (see Figure 4.1). The objective of the simulation was to defend friendly assets against attacks from enemy aircraft. Team performance in this task was indexed by a *team score* which reflected the degree to which the team had achieved the goals of the scenario. Three team goals were employed in this experiment: a) prevent enemy incursion into friendly airspace, b) preserve team assets, and c) protect friendly ground forces (the air base and ranger units). The team score was scaled so it could range from 0–100; a score of 0 indicated that the team had not met any of the goals of the scenario, and a score of 100 indicated that the team had met all goals perfectly. Team score was calculated as the average of the percentages of the three goals that participants had achieved in a trial (i.e., if participants had prevented 70 percent of enemies from entering friendly airspace, preserved 75 percent of team assets, and retained 80 percent of ground forces, their team score for the trial would be 75 percent). Participants received their team score following completion of each trial. As a motivator, teams were compensated for their participation on a sliding scale determined by their average team score.

The air defense task was designed to create moderate intra-team interdependence and collaboration by distributing asset and battlefield information to specific team members, with access to this information dictated by team role. This is a good scenario to study constructs such as trust, since team members needed to rely on their teammates to facilitate optimal team performance. For example, only tanker and strike operators had access to team asset weapon and fuel status information. Therefore, WDs had to rely on their teammates to communicate this resource information to them to adequately perform the task. A variety of performance metrics were automatically captured by the software. Of these, we focused on two performance metrics: team score and average time to prosecute an enemy target.

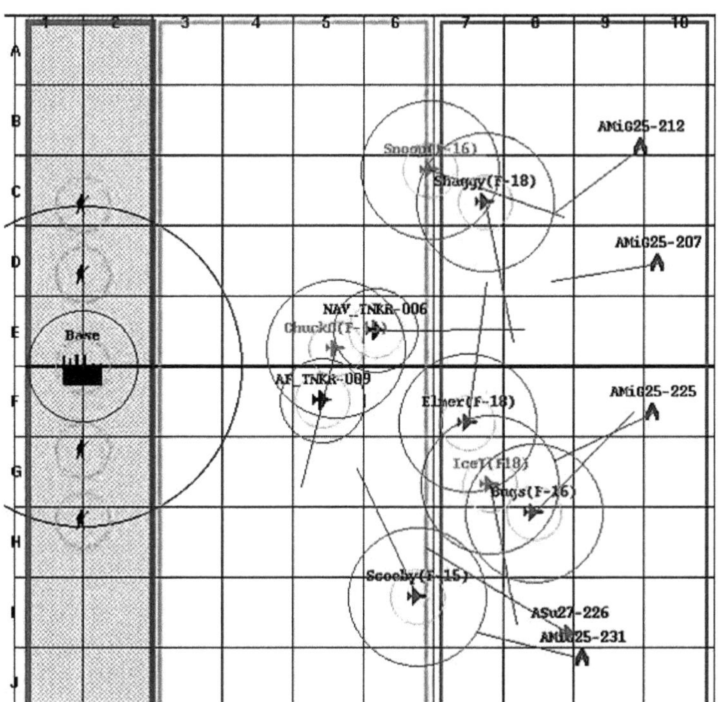


**Figure 4.1 An image from the tactical display of the DDD air defense task**

*Questionnaires* Team communication was assessed using a measure developed for this study and contained 10 items that measured perceived communication and information sharing (e.g., 'If my performance on this task is good, it's because I get the information I need from my teammates'). Reliability was acceptable, with Cronbach's alphas of .77, .83, .92, .82, and .84 for days 1–5, respectively.

Team cohesion was measured using a 13-item scale that is based on a three-dimensional conceptualization of cohesion that includes risk taking (e.g., 'The group avoids saying anything which might upset someone'), attraction to the group (e.g., 'If I were to participate in another group like this one, I would want it to include people who are very similar to the ones in this group'), and instrumental value (e.g., 'The group has influenced me in a lot of positive ways;' Stokes, 1983). Reliability was high, with Cronbach's alphas of .94, .96, .97, .96, and .97 for days 1–5, respectively.

Team trust was assessed with the Organizational Trust Inventory (OTI; Cummings and Bromiley, 1996), which is an 11-item measure of perceptions of trust individuals have regarding other team members and the quality of their teammates' performances (e.g., 'I think other team members met their obligations during the task'). Reliability was high, with Cronbach's alphas of .87, .92, .96, .91, and .91 for days 1–5, respectively.

Collective efficacy (based on Riggs and Knight, 1994) was composed of seven items that assessed each individual's perceptions about the abilities of their team (e.g., 'The members of this team have excellent task skills'). Reliability was acceptable, with Cronbach's alphas of .89, .86, .85, .84, and .86 for days 1–5, respectively.

### *Procedure*

Before selection for the study, potential participants were screened using a biographical survey and the 15-item letter-sets test. These criteria were selected to more closely approximate the personnel requirements of C2 operators. Regarding the biographical survey, to be eligible for the study participants were required to have some college experience (completed at least one quarter or semester) and have some computer experience (enough to develop proficiency with a mouse and keyboard). In addition, participants who indicated that they were color blind were ineligible to participate. As a test of general reasoning ability, participants completed the 15-item letter-sets test (Ekstrom et al., 1976). Items on this test consist of five sets of four letters (e.g., ABCD); test takers are asked to infer the pattern of the letter sets and identify the set that deviates from the pattern rule. Participants were allotted seven minutes to complete the 15 items, with eligibility for this study determined by those scoring nine or more of the 15 items correctly.

The duration of the experiment was approximately 48 hours, conducted across six 8-hour sessions. The first session was devoted to task familiarization, and subsequent sessions to experimental data collection. Upon arrival at the laboratory, participants completed an informed consent document, were introduced to their teammates, and given a brief overview of the experiment. They were then randomly assigned to a team position and received written and oral instructions detailing the DDD simulation, the team's goals, the control of team assets, and the roles and responsibilities of each team position. Following the instructional period, participants were required to complete a proficiency demonstration, administered by the experimenter, which required them to display adequate team role knowledge and ability. Next, teams completed 13 practice trials designed to give them an opportunity to collaborate and strategize under conditions identical to those they would experience during experimental trials. These trials, and all subsequent experimental trials, were 10 minutes in duration.

During experimental data collection sessions, participant experiences varied slightly as a function of team treatment condition. In the permanent team condition, across all data collection sessions, participants resumed their assigned team role (i.e., weapon's director, strike operator, tanker operator) and reviewed a set of reminder instructions detailing the simulation, team goals, and team roles. Next participants completed three 'warm-up' practice trials followed by 16 experimental trials.

Conversely, in the rotation team condition, before beginning each experimental session participants were assigned a new team role according to the schedule displayed in Table 4.1. Participants then reviewed the same reminder instructions provided to permanent condition teams, and completed an identical set of three warm-up trials. However, during these practice trials, rotation team participants were required to again complete a proficiency demonstration, administered by the experimenter, to verify they had retained sufficient team role knowledge and the ability to adequately perform their new team role. Participants who failed to meet this standard were given additional verbal instruction by the experimenter during the practice trials until they were able to do so. Across rotation condition teams, all participants met the required proficiency demonstrations before completion of the practice trials. Following the practice trials, rotation condition teams completed 16 experimental trials. At the conclusion of each day, participants completed the team process questionnaires.

**Table 4.1 Team role rotation schedule**

| | Experimental session | | | | |
|---|---|---|---|---|---|
| **Training Role** | **1** | **2** | **3** | **4** | **5** |
| WD 1 | WD 1 | WD 2 | Strike 2 | Strike 1 | Tanker |
| WD 2 | WD 2 | Strike 2 | Strike 1 | Tanker | WD 1 |
| Strike 1 | Strike 1 | Tanker | WD 1 | WD 2 | Strike 2 |
| Strike 2 | Strike 2 | Strike 1 | Tanker | WD 1 | WD 2 |
| Tanker | Tanker | WD 1 | WD 2 | Strike 2 | Strike 1 |

## Results

### *Questionnaire Data*

Means and standard deviations for all questionnaire variables are presented in Table 4.2. Repeated measures analysis of covariance (ANCOVAs) were used to test for differences between the experimental conditions. ANCOVAs were used to control for team collective efficacy, communication, cohesion, and trust during the first day. Controlling for these variables was critical because the two team treatment conditions were essentially equivalent on the first day of data collection as no team rotation had occurred yet. Unexpectedly, the two groups did not differ on communication, cohesion, or collective efficacy (all *F*'s, *ns*). There were no significant main effects of cross-training, nor was there a significant interaction of cross-training over time when predicting communication, cohesion, or collective efficacy. However, there was a significant data collection day by condition interaction when predicting team trust, $F(3, 81) = 3.58, p < .05$. Unexpectedly, and as illustrated in Figure 4.2, the permanent teams reported higher trust over time relative to the rotation teams.

**Table 4.2 Means and standard deviations of all questionnaire variables by condition**

| | Experimental session | | | | |
|---|---|---|---|---|---|
| **Variable** | **1** | **2** | **3** | **4** | **5** |
| Collective efficacy | | | | | |
| Rotation | 5.01 (.97) | 5.29 (1.30) | 5.82 (1.16) | 5.67 (1.26) | 5.81 (1.27) |
| Permanent | 5.70 (.72) | 5.75 (.59) | 6.05 (.81) | 6.16 (.80) | 5.81 (.92) |
| Team communication | | | | | |
| Rotation | 5.33 (.69) | 5.59 (.99) | 5.77 (1.11) | 5.64 (.92) | 5.78 (1.01) |
| Permanent | 5.80 (.36) | 5.84 (.45) | 5.98 (.69) | 5.87 (.58) | 5.81 (.61) |
| Team cohesion | | | | | |
| Rotation | 5.74 (1.07) | 5.91 (1.02) | 6.02 (1.13) | 5.90 (.92) | 5.99 (.99) |
| Permanent | 5.84 (.79) | 5.70 (.72) | 5.93 (.83) | 5.97 (.73) | 5.91 (.83) |
| Team trust | | | | | |
| Rotation | 5.73 (.92) | 5.94 (.97) | 5.92 (1.10) | 5.66 (1.21) | 5.72 (1.14) |
| Permanent | 6.32 (.61) | 6.02 (.98) | 6.44 (.63) | 6.41 (.50) | 6.24 (.62) |

*Note*: Values in parentheses are standard deviations.

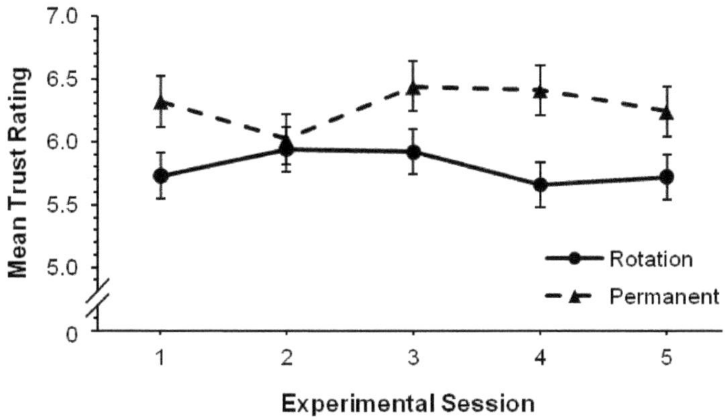


**Figure 4.2 Time by condition interaction when predicting trust. Error bars are standard errors of the mean**

*Team Performance*

During each experimental trial, the DDD software recorded the movements and actions of team and enemy assets. This information was subsequently used to calculate the average team score (computed as previously described) and the average time participants took to prosecute enemy targets in each experimental condition. Prosecution time was calculated as the mean time elapsed from the entry of an enemy target into the simulation to the time it was shot down by team assets. Prior to calculation of team performance averages, the raw reaction time and team score data were transformed to normalize the distributions of each. Team score was recalculated using an arcsine transformation and prosecution time was recalculated using a log transformation.

The team performance data were tested for statistically significant differences between conditions by means of separate 2 (team condition) × 5 (experimental session) mixed analyses of variance (ANOVAs). In all subsequently reported analyses including a repeated measures factor with three or more levels, the Box/Geisser-Greenhouse epsilon correction (e.g., Grieve, 1984) was applied to adjust the degrees of freedom for violations of the sphericity assumption.

*Team score* Analysis of the team score data indicated a statistically significant main effect of experimental session, $F(1.47, 5.88) = 14.10$, $p < .05$. No other sources of variance in the analysis were statistically significant (all $p > .05$). As is illustrated in Figure 4.3, team performance was similar in both team conditions, and increased linearly across experimental sessions.

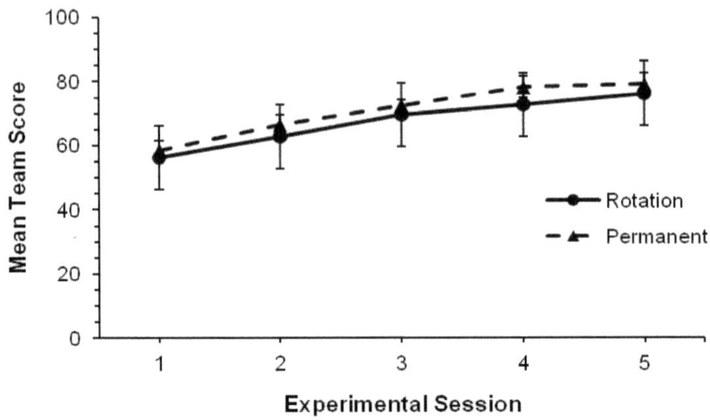


**Figure 4.3** **Mean team score by team condition and experimental session. Error bars are standard errors of the mean**

*Time to prosecute targets* Analysis of the prosecution time data indicated a statistically significant main effect of experimental session, $F(2.43, 9.73) = 14.42$, $p < .05$. No other sources of variance in the analysis were statistically significant (all $p > .05$). As can be seen in Figure 4.4, teams displayed a similar reduction in time to prosecute targets across experimental sessions regardless of team condition.

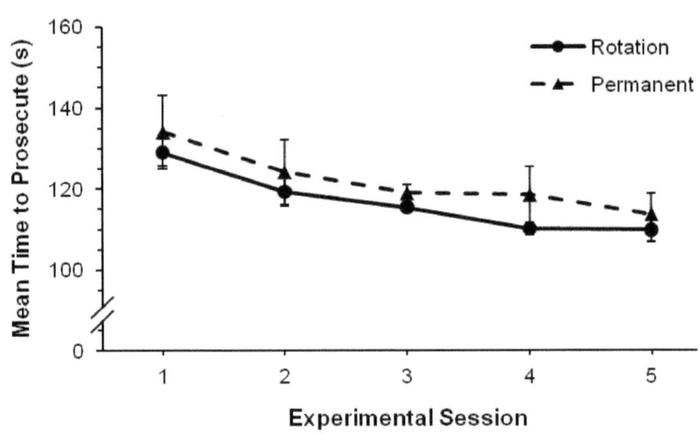


**Figure 4.4** **Mean time to prosecute enemy targets by team condition and experimental session. Error bars are standard errors of the mean**

*Study Correlations*

Presented in Table 4.3 are the correlations among all of the study variables. As expected, many of the psychosocial variables were related to one another. To reduce the impact of common method variance, only the correlations between variables assessed on different days were interpreted. Notably, communication effectiveness was consistently related to higher collective efficacy, higher cohesion, and higher trust the following day. Trust was often related to better communications, higher cohesion, and greater collective efficacy on subsequent days. There were also several instances of collective efficacy and cohesion being related to higher trust.

## Discussion

The purpose of the present study was to explore how cross-training impacts perceptions of communication processes, group cohesion, team trust, and collective efficacy, as well as how cross-training influences team performance. Unexpectedly, cross-training appeared to have little impact on reported collective efficacy, effective team communication processes, and team cohesion. This is inconsistent with some past research that suggested that cross-training facilitates team communication (Volpe et al., 1996). There was, however, a significant effect for team trust, though the effect was counter to predictions. Individuals who were on cross-trained teams reported lower trust over time relative to individuals on permanent teams. While there is a paucity of past research in this area, we did expect that cross-training would promote a shared understanding of task demands and information needs within the team. In addition, and counter to initial predictions, team performance was not improved by cross-training. While this is surprising, it is also consistent with some previous research. For example, McCann and colleagues (2000) found that cross-trained teams evidenced no performance benefits relative to non-cross-trained experimental controls. It is possible that five-person teams are too large to elicit performance benefits from the complex positional rotation cross-training approach employed in the current study.

It appears in the present study that cross-training may actually hinder the development of trust among teammates, at least early in a team's lifecycle. There are several possible explanations for these findings. First, it is conceivable that differences in cross-training methods may have led to differences in experimental outcomes between the current and previous research. Previously reported research indicating that cross-training improves team processes and performance has typically utilized positional modeling (e.g., Cannon-Bowers et al., 1998; Volpe et al., 1996) rather than the positional rotation approach employed in this experiment. It may be that positional modeling provides sufficient additional role and team knowledge to team members to facilitate improved team process development. While rotation may also provide some of these benefits, they may be offset by reductions in role experience as team members rotate through each

**Table 4.3 Correlations among all study variables ($N = 30$)**

| Variable | 1 | 2 | 3 | 4 | 5 | 6 | 7 | 8 | 9 | 10 | 11 | 12 | 13 | 14 | 15 | 16 | 17 | 18 | 19 |
|---|---|---|---|---|---|---|---|---|---|---|---|---|---|---|---|---|---|---|---|
| 1.Efficacy T1 | | | | | | | | | | | | | | | | | | | |
| 2. Efficacy T2 | .05 | | | | | | | | | | | | | | | | | | |
| 3. Efficacy T3 | .17 | .36* | | | | | | | | | | | | | | | | | |
| 4. Efficacy T4 | .34 | .19 | .67** | | | | | | | | | | | | | | | | |
| 5. Efficacy T5 | .02 | .29 | .77** | .59** | | | | | | | | | | | | | | | |
| 6. Comm T1 | .61** | .37* | .41* | .16 | .26 | | | | | | | | | | | | | | |
| 7. Comm T2 | .13 | .50** | .72** | .38* | .66** | .59** | | | | | | | | | | | | | |
| 8. Comm T3 | .01 | .45* | .82** | .45* | .80** | .48** | .82** | | | | | | | | | | | | |
| 9. Comm T4 | .19 | .23 | .59** | .79** | .54** | .17 | .37* | .49** | | | | | | | | | | | |
| 10. Comm T5 | .10 | .30 | .65** | .36 | .70** | .59** | .80** | .84** | .38* | | | | | | | | | | |
| 11. Cohesion T1 | .62** | .13 | .16 | .01 | .08 | .56** | .28 | .14 | .01 | .28 | | | | | | | | | |
| 12. Cohesion T2 | .12 | .36* | .55** | .19 | .49** | .45* | .56** | .61** | .29 | .71** | .58** | | | | | | | | |
| 13. Cohesion T3 | .22 | .33 | .58** | .23 | .52** | .50** | .58** | .64** | .36 | .70** | .56** | .94** | | | | | | | |
| 14. Cohesion T4 | .20 | .32 | .43* | .64** | .30 | .03 | .28 | .34 | .68** | .30 | .21 | .52** | .56** | | | | | | |
| 15. Cohesion T5 | .25 | .44* | .43* | .22 | .29 | .48** | .50** | .46* | .27 | .60** | .63** | .87** | .88** | .62** | | | | | |
| 16. Trust T1 | .66** | .28 | .31 | .14 | .20 | .56** | .22** | .22 | .06 | .15 | .63** | .30** | .34 | .04 | .27 | | | | |
| 17. Trust T2 | .14 | .24 | .55** | .31 | .53** | .31 | .46* | .55** | .21 | .41* | .20 | .33 | .34 | .14 | .27 | .55** | | | |
| 18. Trust T3 | .31 | .34 | .86** | .57** | .69** | .53** | .74** | .84** | .51** | .68** | .23 | .50** | .61** | .42* | .45* | .42* | .66** | | |
| 19. Trust T4 | .27 | .16 | .57** | .69** | .56** | .17 | .36 | .46* | .77** | .30 | .09 | .20 | .29 | .57** | .14 | .29 | .43* | .62** | |
| 20. Trust T5 | .30 | .37* | .78** | .55** | .66** | .44* | .55** | .66** | .61** | .55** | .26 | .56** | .61** | .48** | .42* | .48** | .62** | .82** | .75** |

*Note*: * $p < .05$, ** $p < .01$. Comm = Communications, T1 = Time 1, T2 = Time 2, T3 = Time 3, T4 = Time 4, T5 = Time 5.

team position, inhibiting the maturation of related team processes such as trust. Supporting this perspective, past research has identified ability, integrity, and perceived benevolence as valid indicators of trustworthiness (Colquitt et al., 2007; Mayer et al., 1995). Rotation teams may have had more difficulty establishing behavioral norms, which could have reduced team members' ability to accurately judge trustworthiness. However, as evidenced by the performance metrics, rotation teams performed as well as permanent teams, suggesting that rotation may impair team members' *comprehension* of appropriate teamwork-related performance metrics of individual and team performance, rather than directly undermining task performance. In the current study, rotation team participants changed roles each day, which may have impeded their ability to accurately assess teammates' performances and thus disrupted normal trust calibration among members.

Second, the teams in the current study interacted for a relatively short period of time. Past research has shown that computer-mediated teams develop trust more slowly than face-to-face teams (Wilson, Straus, and McEvily, 2006), and this effect may have been exacerbated by the team rotation manipulation. It may be that rotation teams did not have sufficient time to fully adapt to the computer-mediated nature of communication in this experiment as they changed roles. This explanation is plausible given that individuals on the rotation teams appeared to recover in their trust perceptions to some degree (albeit, not significantly).

Inspection of the correlations between the psychosocial variables examined in this experiment revealed an interesting pattern of relations. Notably, team communication effectiveness was related to higher efficacy, cohesion, and trust, particularly at the fourth and fifth experimental sessions. Past research has highlighted the importance of effective team communication in predicting team performance (Aubert and Kelsey, 2003). Effective communication may help to establish performance monitoring processes, which in turn should facilitate the development of team efficacy perceptions. Communication may also help to foster trust, as has been shown in past research (Jarvenpaa and Leidner, 1999). Trust in turn, appeared to foster effective communications, collective efficacy, and cohesion. Future research is needed to further understand the complex nature of how team processes influence one another.

Counter to initial predictions and previous research (e.g., Cannon-Bowers et al., 1998), cross-training did not facilitate team performance in this experiment. This finding is perhaps not entirely surprising given analogous findings from McCann and colleagues (2000), who used a similar research design (i.e., positional rotation) as well as the observed negative influence cross-training exerted on trust development. As discussed previously, the relationship between cross-training and team performance may actually be moderated by team process factors such as team trust. By negatively influencing the development of trust in this experiment, the beneficial effects of cross-training reported in previous research may have been reduced, causing cross-trained and permanent teams to perform similarly on the air defense task. Further, while the current study suggests that positional rotation may have deleterious effects on initial team process development, its potential benefits may emerge in situations where teammates must flexibly adapt to new team roles and responsibilities (e.g., due to injuries, attrition, etc.; McCann et al., 2000).

### *Limitations and Future Research*

One limitation of the current study was the relatively small sample size, which may have influenced our ability to test team-level relationships. For example, researchers suggest that trust will have the most salient impact on behavior in weak situations (i.e., those where there are few rules to govern one's behavior; Dirks and Ferrin, 2001). This speculation can be tested by looking at the relationship of trust and performance at the team level using sophisticated statistical analyses such as hierarchical linear modeling (HLM; Hox, 2002). However, HLM analyses also require very large sample sizes, which can be prohibitively costly and time consuming. This difficulty is compounded in military domains, where participant availability may be difficult to secure for long durations. A second limitation involves the artificial time periods within which the teams interacted. The teams in the current study interacted over the course of a week, and while this is a very long time in experimental terms, it is actually quite brief for an operational team.

Future research might explore the relationships between team training, processes, and performance using a larger sample size to allow for more varied statistical analyses such as HLM. Further, future research might consider using a combination of laboratory-based teams and actual work teams which interact over a longer time period. Studies that incorporate a longer data collection period may find that rotation teams evidence higher trust, efficacy, cohesion, and communication over the course of a team's lifecycle. It is notable that toward the end of data collection, rotation team members appeared to report higher trust. Perhaps if given enough time, rotation teams would demonstrate better team processes over time. Similar findings have emerged in research exploring the impact of team diversity on performance. Such studies have demonstrated that team diversity initially exerts a negative influence on performance, but over time, diverse teams tend to perform as well as, if not better than, demographically homogenous teams (Jackson, Stone, and Alvarez, 1992).

The results of the current experiment suggest that further research concerning the influence of different cross-training methods on team process development and performance is warranted. Such investigation could help training researchers to refine the appropriate depth of cross-training across a team's lifecycle and maximize its benefits.

## Conclusion

The present study investigated the effects of team cross-training on outcomes such as collective efficacy, communication effectiveness, cohesion, trust, and team performance. There are currently few studies which have empirically evaluated the impact of team cross-training on team process outcomes, particularly studies using a positional rotation manipulation. Cross-training was found to impact the development of team trust but not in the direction originally predicted. Future research is warranted to further explore these effects and to modify training regimes to optimize team effectiveness within the military.

## References

Alberts, D.S. and Hayes, R.E. (2003). *Power to the Edge: Command Control in the Information Age.* CCRP Publication Series.

Aubert, B.A. and Kelsey, B.L. (2003). Further understanding of trust and performance in virtual teams. *Small Group Research, 34*, 575–618.

Bandura, A. (1997). *Self-efficacy: The Exercise of Control.* W.H. Freeman: New York.

Blickensderfer, E., Cannon-Bowers, J.A., and Salas, E. (1998). Cross-training and team performance. In J.A. Cannon-Bowers and E. Salas (eds), *Making Decisions under Stress: Implication for Individual and Team Training* (299–311), American Psychological Association: Washington, DC.

Burke, C.S., Stagl, K.C., Salas, E., Pierce, L., and Kendall, D. (2006). Understanding team adaptation: A conceptual analysis and model. *Journal of Applied Psychology, 91*, 1189–1207.

Cascio, W.F. (1998). *Applied Psychology in Human Resource Management.* Prentice Hall: Upper Saddle River, NJ.

Colquitt, J.A., Scott, B.A., and LePine, J.A. (2007). Trust, trustworthiness, and trust propensity: A meta-analytic test of their unique relationships with risk taking and job performance. *Journal of Applied Psychology, 92*, 909–927.

Cannon-Bowers, J.A., Salas, E., Blickensderfer, E., and Bowers, C.A. (1998). The impact of cross-training and workload on team functioning: A replication and extension of initial findings. *Human Factors, 40*, 92–101.

Cummings, L.L. and Bromiley, P. (1996). The organizational trust inventory (OTI): Development and validation. In R.M. Kramer and T.R. Tyler (eds), *Trust in Organizations: Frontiers of Theory and Research*, (302–330), Sage: Thousand Oaks, CA.

Dirks, K.T. (1999). The effects of interpersonal trust on work group performance. *Journal of Applied Psychology, 84*, 445–455.

Dirks, K.T. and Ferrin, D.L. (2001). The role of trust in organizational settings. *Organizational Science, 12*, 450–467.

Druskat, V.U. and Kayes, D.C. (1999). The antecedents of team competence: Toward a fine-grained model of self-managing team effectiveness. In R. Wagerman (ed.), *Research on Managing Groups and Teams: Context* (vol. 2, pp. 201–231), JAI Press: Stanford, CT.

Ekstrom, R.B., French, J.W., Harman, H.H., and Derman, D. (1976). *Manual for Kit of Factor-referenced Cognitive Tests*. Princeton, NJ: Educational Testing Services.

Finomore, V.S., Knott, B.A., Nelson, W.T., Galster, S.M., and Bolia, R.S. (2007). The effects of multimodal collaboration technology on subjective workload profiles of tactical air battle management teams. *Proceedings of the International Symposium on Aviation Psychology*, *14*, 190–196.

Grieve, A.P. (1984). Tests of sphericity of normal distributions and the analysis of repeated measures designs. *Psychometrika*, *49*, 257–267.

Hox, J. (2002). *Multilevel Analysis: Techniques and Applications*. Lawrence Erlbaum: Mahwah, NJ.

Jackson, S.E., Stone, V.K., and Alvarez, E.B. (1992). Socialization amidst diversity: The impact of demographics on work team oldtimers and newcomers. In L.L. Cummings and B.M. Staw (eds), *Research in Organizational Behavior* (Vol. 15, pp. 45–110), JAI Press: Greenwich, CT.

Jarvenpaa, S.L. and Leidner, D.E. (1999). Communication and trust in global virtual teams. *Organizational Science*, *10*, 791–815.

Johnston, J.H., Cannon-Bowers, J.A., and Salas, E. (1998). Tactical decision making under stress (TADMUS): Mapping a program of research to a real world incident-The USS Vincennes. Paper presented at the RTO HFM Symposium on 'Collaborative crew performance in complex operational systems.' Edinburgh, UK.

Kimmerle, J., Cress, U., and Hesse, F.W. (2007). An interactional perspective on group awareness: Alleviating the information-exchange dilemma (for everybody?). *International Journal of Human-Computer Studies*, *65*, 899–910.

Kozlowski, S.W.J. and Bell, B.S. (2003). Work groups and teams in organizations. In W. Borman and D. Illgen (eds), *Handbook of Psychology: Industrial and Organizational Psychology* (Vol. 12, pp. 333–375), John Wiley and Sons Inc: New York.

Lee, J.D. and See, K.A. (2004). Trust in automation: Designing for appropriate reliance. *Human Factors*, *46*, 50–80.

Levin, D.Z., Whitener, E.M., and Cross, R. (2006). Perceived trustworthiness of knowledge sources: The moderating impact of relationship length. *Journal of Applied Psychology*, *91*, 1163–1171.

MacMillan, J., Entin, E.B., Hess, K.P., and Paley, M.J. (2004). Measuring performance in a scaled world: Lessons learned from the Distributed Dynamic Decisionmaking (DDD) synthetic team task. In S.G. Schiflett, L.R. Elliott, E. Salas, and M.D. Coovert (eds), *Scaled Worlds: Development, Validation, and Applications* (pp. 154–180), Ashgate: Brookfield, VT.

Marks, M.A., Sabella, M.J., Burke, C.S., and Zaccaro, S.J. (2002). The impact of cross-training on team effectiveness. *Journal of Applied Psychology*, *87*, 3–13.

Mayer, R.C., Davis, J.H., and Schoorman, F.D. (1995). An integration model of organizational trust. *Academy of Management Review*, *20*, 709–734.

McCann, C., Baranski, J.V., Thompson, M.M., and Pigeau, R.A. (2000). On the utility of experiential cross-training for team decision making under time stress. *Ergonomics*, *43*, 1095–1140.

Mullen, B. and Cooper, C. (1994). The relation between group cohesiveness and performance: An integration. *Psychological Bulletin*, *115*, 210–227.

Parasuraman, R. and Riley, V. (1997). Humans and automation: Use, misuse, disuse, abuse. *Human Factors*, *39*, 230–253.

Riggs, M.L. and Knight, P.A. (1994). The impact of perceived group success-failure on motivational beliefs and attitudes: A causal model. *Journal of Applied Psychology*, *79*, 755–766.

Salas, E., Nichols, D.R., and Driskell, J.E. (2007). Testing three team training strategies in intact teams. *Small Group Research*, *38*, 471–488.

Salas, E., Cooke, N.J., and Rosen, M.A. (2008). On teams, teamwork, and team performance: Discoveries and developments. *Human Factors*, *50*, 540–547.

Stokes, J.P. (1983). Components of group cohesion: Intermember attraction, instrumental value, and risk taking. *Small Group Behavior*, *14*, 163–173.

VandeWalle, D., Cron, W.L., and Slocum, J.W. (2001). The role of goal orientation following performance feedback. *Journal of Applied Psychology*, *86*, 629–640.

Volpe, C.E., Cannon-Bowers, J.A., and Salas, E. (1996). The impact of cross-training on team functioning: An empirical investigation. *Human Factors*, *38*, 87–100.

Walumbwa, F.O., Wang, P., Lawler, J.J., and Shi, K. (2004). The role of collective efficacy in the relations between transformational leadership and work outcomes. *Journal of Occupational and Organizational Psychology*, *77*, 515–530.

Wilson, J.M., Straus, S.G., and McEvily, B. (2006). All in due time: The development of trust in computer-mediated and face-to-face teams. *Organization Behavior and Human Decision Processes*, *99*, 16–33.

# SECTION TWO
# Multinational Challenges for Trust

Chapter 5

# Trust in Swift Starting Action Teams: Critical Considerations

Jessica L. Wildman,[1,2] Stephen M. Fiore,[1,3] C. Shawn Burke[1] and Eduardo Salas[1,2]
*Institute for Simulation and Training,[1] Department of Psychology[2] and Department of Philosophy,[3] University of Central Florida, Orlando, FL, USA*

Sena Garven
*US Army Research Institute, Ft. Leavenworth Research Unit, Fort Leavenworth, KS, USA*

## Introduction

Teamwork has long been an important area of study for the military (e.g., Campbell and Campbell, 2009; Deluga, 1995; Gurbuz, 2009). As international military work becomes more collaborative, and as the rapidly changing environment forces teams to become more adaptive as well, many of these teams are being formed and disbanded quickly to solve complex problems. For example, provincial reconstruction teams are composed of military units embedded with civilian diplomatic agencies and they often are pulled together quickly to address pressing humanitarian needs. Military transition teams are composed of both US military and host national military members, and they must immediately begin their work together to get the host national forces trained while managing the complexities of working in political instability. These types of teams encounter particularly difficult challenges for performance given how little time they have to prepare for their work before engaging in critical, fast-paced missions.

In all of these situations, relatively well-trained individuals are brought together to quickly address some demanding, complex problem, and the involved parties often have little to no prior knowledge of one another. This type of team can be referred to as a swift starting action team (STAT; McKinney, Barker, Davis, and Smith, 2005). STATs are a subset of action teams (i.e., teams that engage in performance events that require improvisation in unpredictable circumstances; Sundstrom McIntyre, Halfhill, and Richards, 2000) that are characterized by the following hallmarks: (1) comprised of well-trained strangers, (2) almost immediate performance, and (3) high levels of risk (McKinney et al., 2005). Although only recently receiving attention in organizational literature, STATs have been prevalent in the field for many years, especially in the military. Certainly, the complex and risky nature of such teams requires a precise understanding of their functioning and the factors which may aid or impede their successful performance.

In these STAT contexts, having appropriately calibrated levels of trust between team members plays a critical role in the success of their endeavors. For example, if the members within a provincial reconstruction team do not trust one another enough, the team members may be spending an inordinate amount of time double-checking one another's work and engaging in task-related arguments, which can lead to wasted cognitive capacity, creating more tension and hostility between the two forces and leading to more severe issues such as betrayal or violence. Alternatively, if the team members trust one another too much and stop monitoring each other and re-evaluating trustworthiness, individuals could become complacent and begin making mistakes that are not caught because there are low levels of monitoring and backup behavior occurring. Therefore, the accurate calibration of trust in STAT contexts in particular is a highly relevant and timely topic for the military.

While the prevalence of STATs seems to be increasing, little research exists to explain their unique processes and functioning. This is especially true of the research on trust in STATs. McKinney, Barker, Smith, and Davis, (2004) have investigated the role of communication in STATs, finding that the early moments of team interaction are often the defining factors of subsequent team interactions. However, while this initial research of STATs recognizes that trust is an important consideration in these types of teams, no theory exists as to how trust is calibrated in such environments. Therefore, the primary purpose of this chapter is to highlight the unique influences on trust calibration in military STAT contexts. We begin by reviewing and critiquing the extant trust literature with a specific focus on the deficiencies for describing trust development in these particular types of teams. We then describe the most critical stressors and team factors to consider when examining trust in STAT-type military teams. Finally, we conclude with a discussion of the practical implications of these considerations in terms of training of military STAT members.

## Review of Extant Stat Relevant Trust Literature

There is a vast amount of theoretical and empirical research examining trust in a variety of contexts. Several well-known definitions of trust have been put forth (see Table 5.1). Several clear themes emerge across all of these definitions. First, trust is a psychological state held by an individual. Second, this psychological state involves expectations about another person's conduct. Third, this psychological state involves feelings of vulnerability and interpersonal risk. In the remainder of this chapter, we adopt an integrated understanding of trust as a psychological attitudinal state held by an individual about another person or persons that involves expectations and feelings of vulnerability and risk.

The majority of the research examining interpersonal trust development and trust in teams adopts the theory of trust put forward by Mayer, Davis, and Schoorman

**Table 5.1 Definitions of trust**

| Definition | Source |
|---|---|
| Willingness of a party to be vulnerable to the actions of another party based on the expectations that the other will perform a particular action important to the trustor | Mayer et al., 1995 |
| A psychological state comprising the intention to accept vulnerability based upon positive expectations of the intentions or behavior of another | Rousseau et al., 1998 |
| Confident positive expectations regarding another's conduct | Lewicki et al., 1998 |
| The expectancy of positive outcomes based on the expected actions of another party in an interaction based on uncertainty | Bhattacharya et al., 1998 |
| The confidence an individual has in a partner's willingness to be responsive to a person's needs, even when they conflict with the partner's own preferences | Rempel et al., 2001 |

(1995). This theory posits that trust attitudes are formed based on assessments of three characteristics of the trustee by the trustor: *ability*, *benevolence*, and *integrity*. Ability is the perception of the skills, competencies, and abilities of the trustee that would lead the trustor to believe that they are capable of performing well in some domain. Benevolence is the perception that the trustor that the trustee truly desires to do good toward the trustor. It is the extent to which the trustor believes the trustee will behave in the best interest of others, not at others' expense or putting their own needs ahead of the group. Finally, integrity is the extent to which a trustee is seen as being honest, dependable, and moral. An individual with high integrity would always try to do the right thing, no matter the consequences or whether anyone would ever know. Mayer and colleagues (1995) suggest that these three characteristics are combined to predict a person's willingness to be vulnerable to that trustee. Higher perceived ability, benevolence, and integrity in a trustee should result in higher trust attitudes aimed at that trustee.

While research has found ample support for this model of interpersonal trust development (e.g., Colquitt, Scott, and LePine, 2007), this particular theory has limitations when applied to the STAT context. Specifically, this theory was developed for contexts in which there is ample time for individuals to make assessments of the three trustee personality characteristics before forming trust attitudes. Internal traits such as ability, benevolence, and integrity can only truly be assessed after some substantial amount of interaction with another individual has occurred and judgments of those characteristics can be made based on their behavior. In a situation in which several individuals have just met and must immediately begin working with one another, such as in STATs, it will be nearly impossible for team members to determine with any level of confidence, let alone accuracy, one another's ability, benevolence, and integrity. Therefore, this particular model of trust development does not provide enough information regarding what other cues and characteristics could potentially influence trust attitude formation when individuals first meet and do not have enough time to engage in any form of assessment of teammates.

Encouragingly, one theory of trust development has begun to tackle this exact issue. Meyerson, Weick, and Kramer (1996) suggest that there is a unique form of trust, known as swift trust, which develops in temporary teams that have little time for socialization. This theory of trust development addresses several of the limitations of the mainstream trust literature. First, they acknowledge that in groups composed of individuals with diverse skills and limited history together, there is very little time to develop trust attitudes in the traditionally expected manner. Second, because of this, they suggest that individuals 'import' their trust attitudes from other settings. Their predispositions to trust, their categorical assumptions regarding different types of people, and their implicit theories about trust all influence the formation of initial trust in a temporary setting.

Our understanding of trust in STATs follows much the same logic: we assume initial trust attitudes are formed based on the immediately salient surface-level characteristics of the individuals within the team, combined with the pre-existing knowledge and affective reactions that each person has associated with those characteristics. In other words, before an individual has a chance to discover the ability, benevolence, and integrity of their fellow teammates, they will make initial 'gut' trust judgments based on visible characteristics and the associated prior experiences and stereotypes.

While Meyerson and colleagues' (1996) theory of swift trust takes several important steps towards more accurately describing trust development in STAT-type contexts, there are still several critical considerations they do not fully address. They do not focus very heavily on the influence of stressors in the team environment on the development of trust attitudes. They also do not highlight the potential influences of team diversity and power differences among team members. In this chapter, we elaborate on this work in order to illuminate how such theorizing can be expanded via consideration of a set of important contextual factors that are part of the military operational environment. We show how those factors pertinent to the STAT environment can influence the formation of trust attitudes among team members. In the following sections, we describe in detail several environmental variables that STATs are likely to encounter and how each of these influences could change the way trust attitudes are formed within the team (Table 5.2).

**Table 5.2 Critical considerations for trust in STATs**

| Consideration | Key issues for trust calibration |
|---|---|
| Time pressure | • Can lead team member to use limited information and simple heuristics (e.g., stereotypes) to calibrate levels of trust<br>• Can make team member less sensitive to relevant and accurate trustworthiness cues<br>• Can cause affective reactions which skew the calibration of trust towards others |
| Perceived risk | • Can cause the team to be predisposed to lower calibration of trust due to high consequences of failure<br>• Extremely low risk can also lead to low levels of trust calibration because trust is deemed unnecessary |
| Diversity | • Can lead to the use of stereotypes and biases formed about diverse populations for the calibration of trust<br>• Can lead to in-group/out-group bias and favoritism in terms of calibration of trust<br>• Can lead to the fundamental attribution error influencing the attribution of trust-related behaviors to either external or internal influences depending on in-group or out-group status |
| Power | • Can lead to imbalance in trust calibration related to differences in power among team members<br>• Can lead to inaccurate trust calibration based on certain types of power that may not be indicative of trustworthiness |

## Time Pressure and Trust in STATs

The very definition of STATs implies that time pressure is an inherent challenge within these types of teams. Due to the short time between STAT formation and deployment there is time pressure with respect to calibrating the level of trust to have for members within the team. Members of STATs are often very aware of the task deadline, and they believe that there is a cost or reward associated with timely performance (Ordóñez and Benson, 1997). As such, they are cognizant of the need to make decisions in the absence of being able to reflect upon critical factors that would influence their decisions. Therefore, the presence of such time pressure can have an impact on the calibration of trust attitudes toward fellow team members.

While time pressure has been shown to have many effects on cognitive processing, affective reactions, and the corresponding behavior, we focus on those most relevant within a STAT framework. In terms of cognitive processing, time pressure has been shown to impact whether there is a primary focus on decision quality (Van der Kleij, Lijkwan, Rasker, and De Dreu, 2009; Payne, Bettman, and Johnson, 1995). Under high levels of time pressure, team members tend to use a combination of the following activities: limit their search for information (Adelman, Miller, Henderson, and Schoelles, 2003); use less rigorous and complicated decision-making strategies (e.g., simplify the problem, use heuristics, tend not to consider multiple alternatives, Van der Kleij et al., 2009; Chu and Spires, 2001; Zakay, 1993); narrow focus to a limited range of task-salient cues (Van der Kleij et al., 2009); important information is often discounted, and negative information that gains attention is often given greater weight (Edland, 1994); and leaders who accelerate their cognitive processing send less information to team members (Adelman et al., 2003).

These adaptations in cognitive processing will have a serious impact on the formation of calibrated trust, as members may not pick up on those cues or characteristics of fellow team members that would assist in accurate trust calibration. Or, if they can pick up on such cues, they may only attend to a limited amount and/or give greater weight to those cues which reinforce preconceived stereotypes. For example, team members may discount positive information about the past performance of a fellow team member based on the regalia that indicates they won an award for bravery, while instead giving more weight to the negative rumors they heard from a third party that this individual makes a lot of mistakes, causing them to use incomplete evidence when calibrating their level of trust for that team member.

Conversely, Adelman et al. (2003) reported that team members who adapted to time pressure (i.e., were able to maintain performance levels under pressure) were proactive in their communication, often relying on tenets of implicit coordination to know when those members with whom they were interdependent needed information. While it could be argued that perhaps members who are aware of the important trust cues could work to relay this information to those with whom they are most interdependent, often, with STATs, the shared mental models upon which proactive communication relies, do not yet exist. In other words, because of the limited history between team members in STATs, relying on implicit coordination to relay trust information will be more challenging than in other types of teams.

In addition to impacting cognitive processing, time pressure also has a significant impact on an individual's affect about the task. For example, research has shown that individuals making a decision under a strict deadline exhibit higher levels of anxiety and energy toward the task than individuals making a decision with no deadline (Maule, Hockey, and Bdzola, 2000). Specifically, research has also shown that time pressure often causes the task to be seen as being less pleasant and enjoyable (Isenberg, 1981). These negative affective states such as anxiety and boredom may inadvertently color team members' trust attitudes in a negative way, causing them to have less trust in others than they would if they were not feeling these negative emotions. This, in turn, may also contribute to the short-cuts taken within cognitive processing, as individuals are not motivated to invest in the task due to its perceived unpleasantness. This, too, would attenuate the quality of the trust calibration in that members may short-circuit the already limited time they have. Finally, behaviorally, individuals operating under time pressure have been shown to become more task focused and less relationship focused. This could potentially lead to STAT members focusing heavily on task cues instead of trust-relevant personal cues, which would be minimal in the early stages of STAT, to the decrement of accurate trust calibration within the team.

## Perceived Risk and Trust in STATs

Deutsch argued that trust comes about on the backdrop of a risky situation: 'Risk-taking and trusting behavior are thus really different sides of the same coin' (1958, p. 266). Bachmann states that trust 'produces risk, as it is inevitable that a social actor who decides to trust another actor extrapolates on limited available information about the future behavior of this actor' (2006, p. 451). As it appears to play an important role in trust development, it is therefore necessary to clarify what is meant by risk. Risk can be defined as 'things, forces, or circumstances that pose danger to people or to what they value' (Stern and Fineberg, 1996, p. 215). The perception of risk can further be described as a trustor's belief regarding risks in a situation outside of the considerations involving a relationship with a particular trustee (Mayer et al., 1995). The level of risk in a situation has a huge bearing on whether or not trust is even necessary (Chiles and McMackin, 1996). Without some level of risk, there is no need to trust (Dasgupta, 1988). STATs like paramedical, aviation, or military teams often face the risk of dying or not saving the lives of others if their mission fails. Therefore, by nature, trust is even more relevant and necessary in STATs than in any other type of team. When there is more at stake, there is an even higher need to accurately calibrate trust levels. Therefore, in order to reach the appropriate calibration of trust within the team, the relevant cues regarding the trustworthiness of the team members would need to be made salient quickly in order to counteract the negative influences of perceived risk.

One view of trust and risk suggests that trust is a substitute for control in a relationship, and therefore perceptions of risk within the team may be minimized by increasing the level of regulation or control in a relationship (Aulakh, Kotabe, and Sahay, 1997). If a trustor believes he/she can control the behaviors of a trustee, perceptions of risk in terms of engaging in a relationship with that trustee decline. For example, in a deployed team consisting of both military and civilians, the military members may believe they have control over the civilian members, and therefore may trust them more, at least regarding task completion, than they would if they did not believe they had control. However, this all depends on the particular context under which the team members are interacting. If the team members have the ability to closely monitor one another's work and therefore reduce the risk of mistakes or failure, trust may be calibrated at higher levels than if monitoring is difficult.

However, while there is little research in this area, it should be noted that low perceived risk may not necessarily always result in high trust. This could be explained by two different mechanisms. First, even in situations characterized by low risk, the trustor cannot be immediately certain that a trustee is entirely trustworthy. Therefore, an individual may perceive certain cues as indicating a low level of trustworthiness in another's intentions or competence regardless of the level of risk in the situation. This may happen, for instance, when individuals hold particularly strong negative stereotypes toward a certain type of person. For example, in a military situation, this could be assuming technical proficiency (or lack thereof) due to one's specialty. Second, when there is little to no risk inherent in a situation, the prospect of failure is no longer an issue to the team members, and therefore trust may simply not be necessary for effective interaction within the team (Chiles and McMackin, 1996). In other words, if there is no risk associated with making a mistake or betraying the team, it doesn't matter if the team members trust one another because breaking that trust will have no consequences. At the other extreme, under highly consequential circumstances, trust may also be irrelevant because there is a mission that must be completed and there is no option but to assume a minimal level of competence for each team member. Of course, in the STAT context, the likelihood of having a situation with no risk is very slim, and therefore risk should always be considered as an influence on the calibration of trust in STATs. Because there are so many barriers to the accurate calibration of trust in STAT contexts, it is important to keep in mind how risk can influence the development of trust as well as how trust can help the team to engage in necessary risk-taking behavior.

## Diversity and Trust in STATs

Another characteristic of the STAT environment that can have a real influence on trust development is the diversity of the team. Diversity refers to 'the differences (or similarities) among members of some particular collectivity' (McGrath, Berdahl, and Arrow, 2000, p. 22). McGrath and colleagues (1995) suggest that there are five clusters of attributes that make a team homogenous or diverse, (a) demographics, (b) task-related knowledge, skills, and abilities, (c) values, beliefs, and attitudes, (d) personality and cognitive and behavioral styles, and (e) status in the works groups' embedding organization. As previously mentioned, military STATs such as military transition teams or provincial reconstruction teams often comprise a mix of civilian and military members and/or multiple nationalities. These differences in cultural history, values, beliefs, and knowledge can lead to difficulties in trust development. Certain diversity attributes may have a very significant impact on trust development in STAT contexts due to their outwardly observable nature.

Demographic (e.g., race, gender, age) and status (e.g., rank, affiliation) attributes are usually easier to perceive early on in a relationship as they are external to a person. In STAT contexts team members will have very little time for socialization before beginning their work, and may, therefore, rely heavily on these surface-level characteristics when forming trust attitudes. For example, imagine that group member A can see that group member B is from a particular training background (e.g., combat arms) and has attained a particular rank. Past experiences with similar individuals may trigger pre-existing beliefs or stereotypes about that category of soldier. It is possible they may have been on a team where an individual with those characteristics was a low performer and repeatedly hindered the team's success, or they may have had a positive experience with a similar individual in which that person performed heroically. These stereotypes and biases will likely lead group member A to automatically form a particular level of trust for group member B immediately upon meeting them, despite the fact that they actually have no true knowledge of this new person's personality, background, or intentions.

The influence of stereotypes and biases is likely to be much more pronounced in highly diverse teams where very different demographics are coming together and in-group/out-group biases (Tajfel, 1982) are triggered. Individuals naturally tend to gravitate towards individuals that they consider to be part of their in-group, and to shy away from individuals considered to be part of their out-group. Research has shown that individuals engage in generally more positive behaviors toward members of their in-group such as offering help and sharing resources (Tajfel, 1982). In these highly diverse STATs, it is likely that individuals that come from similar groups external to the team (e.g., similar training background, prior units served in) will 'band together' and trust their own in-group members more, causing factions or faultlines that could drive the team apart. This could be detrimental to the overall success of the team, especially if the team task is highly interdependent and requires a similar level of sufficient trust among all team members.

Finally, when teams are formed and there is a demand to perform immediately (i.e., high time pressure/stress), attentional narrowing may cause team member attention and effort to be primarily directed towards quickly figuring out how to reach a particular goal rather than on trying to assess the trustworthiness of their teammates (Easterbrook, 1959; Kohn, 1954). This may attenuate the effects of diversity attributes until task-related or interpersonal issues appear, at which point members' perceptions of each other will become more salient. In other words, because the team members will be devoting the majority of their cognitive resources to the task, they may be less likely to notice trust-related cues and activate stereotypes and biases in order to calibrate trust attitudes. However, the opposite could also be argued: if team members are low on cognitive resources due to the demanding nature of the task, the team members may rely more heavily on their own culturally derived understandings of trust and preconceived stereotypes about in-groups and out-groups, instead of assessing more relevant cues that indicate the actual trustworthiness of their fellow team members. Obviously, research is needed to parse these seemingly contradictory hypotheses.

While discussing how diversity may impact trust development in STATs we should consider the fundamental attribution error. This is the tendency of people to attribute another's behavior to dispositional factors while the effects of their own behavior are excused by situational factors (Gilbert and Malone, 1995). In other words, when team member A makes a mistake in a team, they will likely attribute their own mistake to situational causes (i.e., causes outside of their control) such as receiving incorrect information or some sort of technical difficulty. But when team member B makes a mistake, team member A will attribute team member B's mistake to incompetence or other characteristics more likely to be within an individual's ability to control. This attribution error may evolve into a stereotype regarding the competence of individuals in that particular demographic group and may carry over to other trust situations in the future. Overall, it is critical to consider the influence of diversity on the formation of trust attitudes in STATs.

## Power and Trust in STATs

The various types and levels of power within a STAT could potentially influence the way that trust develops between team members. One of the classic theories of power developed by French and Raven (1959) suggests that social power comes from five different sources: *expert*, *referent*, *legitimate*, *coercive*, and *reward* power. Expert power comes from an individual's skills or abilities in an area that others need. Referent power stems from feelings of identification and liking for an individual. Legitimate power comes from an official rank or position that puts an individual in power. Coercive power, in which force is used to make others submit, is a serious potential issue in military settings. Finally, reward power stems from the ability of an individual to provide desirable outcomes to others. Frost and colleagues (1978) found that power, measured as the influence a person has as rated by others, was strongly correlated with trust. In other words, highly influential people, regardless of the source of their power, were more likely to be trusted than less influential people. This can often been seen in the high levels of trust team members have for their leaders, or the high levels of trust citizens have in their political party or government (although it is important to note that this is not always the case).

Several of these types of power could have significant effects on the calibration of trust in STAT contexts. For example, in a provincial reconstruction team, there are often civilian members working directly with military members. It is possible that the civilian members will perceive themselves to be at a lower level of legitimate power than the military members, given the governmental nature of the military and their ability to use force to achieve their goals. Based on this assumption of power, the civilian members immediately develop an unwarranted amount of trust in the competence of the military members simply based on the legitimate power associated with their position. Simultaneously, the military members may also perceive themselves as having higher levels of legitimate power. This may cause the military members to assume they are more capable and worthy of their position on the team, and they may tend not to trust the civilian members with responsibility as much as their own in-group of military members. Conversely, the civilian team members may feel as if the military members are quite capable, but are corrupt or violent (based on stereotypes of the military), and therefore their capabilities should be trusted but their motives should not be trusted. This could lead to a rift in the team between civilian and military members. In order to avoid this potential issue, the STAT leader may emphasize the important role that each team member plays regardless of their civilian or military status and try to downplay the status or affiliation differences between them.

Specifically for STATs, legitimate power is likely to be the most salient type of power present within the team initially. As mentioned previously, because the team members will have very little time to explore one another's deep-level characteristics such as ability or integrity, judgments of power, like trust, will be based on surface-level characteristics. They will likely be looking for symbols that indicate each team member's status and rank, such as uniforms and rank insignia if they are familiar with the meaning of these symbols. These symbols will serve as indicators of legitimate power, and the team members may determine their trust attitudes based on their understanding of each team member's power and status within the team. Conversely, some civilian team members may enter the team with preconceived stereotypes of soldiers as being uneducated and rash, and ignore the potentially relevant cues that indicate their competence. However, as the team works together and each team member's skills and abilities become more apparent based on behavior instead of stereotypes or insignia, expert power may become more influential on the readjustment of trust levels.

Expert power is very different from legitimate power. It may be that within a particular team, there is a formally appointed leader who technically holds the legitimate power and is most trusted in the beginning of team formation. However, this does not necessarily mean that the leader also holds the expert power. There may be a different team member who is highly skilled, and over time as this becomes apparent to the other team members, that person may hold the most expert power in the team. It may be that the trust among the team members grows strongest toward the team member with the expert power rather than toward the formally appointed leader after enough time has passed to make the ability of the team members more salient. While this may not necessarily be positive or negative in terms of its effect on performance, it is important to understand that differences in power between team members will change the pattern, rate, and strength of trust development and will have different influences over time within STATs.

## Training Recommendations to Enhance Trust Calibration in STATs

In this section we revisit the critical considerations for trust development so as to discuss their implications for training to prepare for being a member of a STAT (Table 5.3). This is a necessarily hypothetical discussion in that little research has explored these factors in such a specific context. As such we offer these as items for consideration by the operational community. Note that these training strategies are meant to attenuate the overall impact that these factors have on STATs. Given the uncertain characteristics of STAT tasks, we focus on more general strategies that can prepare team members for the various environments in which they will be operating. As such, their impact on calibration of trust is more distal as these are meant to address the outcomes of situations where trust formation is hindered.

Time pressure is inherent in the contexts in which STATs operate. Similarly, many STATs operate under conditions of high perceived risk. Given this, training strategies should focus on mitigating the consequences of trust miscalibration on team member processes and performance in conditions of time pressure and risk. First, STATs members need to understand the impact time pressure and risk have on trust calibration as well as on team processes and performance. As noted, those under time pressure engage in a narrow search for information and tend to use heuristics in decision making. These heuristic cues may not be appropriate for the problem context and thus may not be relevant in calibrating the trustworthiness of their fellow team members. Along these lines, those perceiving high levels of time pressure in the situation may become overwhelmed and become cognitively narrowed onto a particular task-related aspect of the mission, and, for example, not execute team behaviors such as asking for, or providing, backup.

Cognitively based training may be effective in this context given that it could be designed to both increase trainees' awareness of the consequences of time pressure and risk on trust calibration while providing recommendations for overcoming these problems. Current military training can be adapted to incorporate an awareness of how time pressure and perceived risk levels alter cognition and the development of trust attitudes. Trainers in situations where trainees are swapped in and out, thus forming teams of varied compositions, could

**Table 5.3 Training recommendations to enhance trust calibration in STATs**

| Consideration | Training recommendations |
|---|---|
| Time pressure | • *Scenario-based training*: designed to mimic the types of time-pressured situations that STAT members face and provide feedback regarding accurate trust calibration within that situation |
| Perceived risk | • *Stress exposure training*: team members are exposed to various stressors that could cause team members to inaccurately calibrate trust and they are given strategies to cope with these stressors |
| Diversity | • *Role-playing training*: team members engage in scripted scenarios with fellow team members in which the consequences of biases and stereotypes are made salient |
| Power | • *Assertiveness training*: team members are taught to focus on their unique contributions and use strategies to present their knowledge within the team |

make explicit the impact of time pressure and the trainees' perceived risk on all decisions made, including calibrating trust of fellow team members. In addition to noting the impact of time and risk, trainers could also make recommendations for overcoming these problems such as giving feedback that illustrates the errors in trust calibration that arise when information search for trust cues is narrow and/or faulty heuristics are used. We turn next to a set of methods that could be used to deliver cognitively based training.

If a separate line of instruction is warranted and possible rather than modifying existing training, a stand-alone 'scenario-based training' program (e.g., Oser, Gualtieri, Cannon-Bowers, and Salas, 1999; Salas and Cannon-Bowers, 2000) can be developed to mimic the types of situations STAT members face. Here the training would be structured such that the training environment provides multiple cues and these are differentially relevant to the decision-making context and to determining the appropriate levels of trust for the team. Trainees would then have a limited amount of time to choose from the cues and make a decision regarding whom to trust with what tasks. Further, the training would be structured such that the cues need to be integrated in order to make a decision. This would address the problematic use of simplified decision heuristics in time-pressured situations. Feedback tailored to the trainee's performance would then highlight any errors which arise when information search was poor and/or cue integration was in error.

In order to address problems in STATs associated with 'perceived risk,' training should similarly target the consequences of this experience. In particular, because risk produces stress and anxiety in team members, training which familiarizes team members with stress in complex decision contexts may ameliorate any negative outcomes of this stress on trust calibration and team interaction. This could include 'stress exposure training' (e.g., Driskell and Johnston, 1998) where team members are provided with knowledge about the stressors of the operational context and with strategies devised to help them cope with these stressors. Similarly, 'stress inoculation' training (e.g., O'Donohue, Hayes, and Fisher, 2003) works to provide not only coping skills, but also stress avoidance skills.

Much military training and mission preparation guidance already incorporates the concepts in the previous paragraph, thus it would be a small matter to include the team-trust aspects as well. The consequence of team diversity in STATs manifests itself primarily in the implementation of stereotypes or implicit biases team members may hold about what kinds of people are trustworthy. In particular, as with the aforementioned characteristics, team members are likely to take short-cuts to decision making when finding themselves with limited time and limited information. Power differences can also produce situations where team members engage in attitudes and behaviors not conducive to optimizing team performance. For this reason, training needs to make team members aware of these biases and problems, and provide strategies to attenuate their impact.

First, 'role-playing' training (e.g., Salas, Burke, and Cannon-Bowers, 2002) could be implemented in order to provide team members with first-hand experience of the consequences of biases and stereotypes on trust calibration which may arise in situations of team diversity. Role-playing in the context of STATs might involve the development of scripted scenarios, with team members taking on the part of those from different groups. This is followed by some form of structured discussion where the experiences are drawn out of team members in order to make explicit their experiences. A more specific form of role-playing, called 'multi-cultural' training (e.g., Hayes et al., 2004), could be implemented if team members are likely to experience STAT contexts involving much diversity in culture. Here the training is designed to sensitize the team member to the impact of cultural biases by attending the consequent affective reactions (e.g., negative emotions) and trust-related behavioral outcomes (e.g., ostracism).

Second, 'assertiveness training' (e.g., Smith-Jentsch, Salas, and Baker, 1996) might be useful to address any perceived problems of power distance. This form of training focuses the team members on their unique contributions to the team and provides them with strategies which help them to more forcefully present that knowledge in the course of teamwork. As an example, members of a provincial reconstruction team could be trained in situations where civilians who have much-needed knowledge are taught to assertively present their input even when such members have lower legitimate power than military members. This could simultaneously address the problem where military members perceive that they have a higher level of legitimate power. Essentially, all members of a team would come to understand that role input can sometimes override power structures, leading to more accurate calibration of trust among the team members.

## Implications and Conclusions for Military STATs

Trust calibration in a military STAT is currently a very relevant topic for service members of all ranks and grades as well as for their civilian counterparts who are more frequently being asked to serve alongside them in harm's way. While the preferred method of team development is formation, then training together, followed by mission assignment and completion, in today's world that is not always the case. All too often, teams are formed within hours or even minutes of needing to perform a mission. The likelihood of successfully completing the immediate mission improves if team members are able to calibrate trust among team members and within the team as a whole. In the current vernacular, trust is discussed in terms of developing it. In this chapter, we would like to propose that, more specifically, trust levels are calibrated in that team members become aware of the perceived levels of ability (competence), benevolence, and integrity of other team members and adjust their own behaviors accordingly. Adequate calibration of trust is vital to team performance, as team members who overestimate trust levels in another team member risk something 'falling through the cracks' and team members who under-calibrate trust may perform unnecessary backup behaviors.

Both types of errors take up valuable time and cognitive capacity better used for mission accomplishment.

The training recommendations in this chapter leverage the repetitive conditioning training the military already uses for training operating and decision making under pressure as well as the after-action review used for incorporating the lessons learned into the next round of training or the next mission. The idea of trust calibration and the factors that influence it could be easily incorporated into the current programs of instruction where service members train in large groups which interchange members for smaller training teams. The recommendations in this chapter are based on both empirical research as well as untested, but reasonable, theory. In the future, we hope to be able to more fully test our implied training models and provide more complete, valid, and usable guidelines for trust calibration to service members.

### Acknowledgement

This work was partly supported by the Mirum Corporation through Army Research Institute SBIR: ARI-SBIR A04-T031. The views expressed are those of the authors and do not necessarily reflect the opinion of the Mirum Corporation or ARI.

### References

Adelman, L., Miller, S., Henderson, D., and Schoelles, M. (2003). Using Brunswikian theory and a longitudinal design to study how hierarchical teams adapt to increasing levels of time pressure. *Acta Psychologica*, *112*(2), 181.

Aulakh, P.S., Kotabe, M., and Sahay, A. (1997). Trust and performance in cross-border marketing partnerships. In P.W. Beamish and J.P. Killing (eds), *Cooperative Strategies: Vol. 1. North American Perspectives* (pp. 163–196). San Francisco: New Lexington Press.

Bachmann, R. (2006). Trust, power, and control in trans-organizational relations. In R.M. Kramer (ed.), *Organizational Trust: A Reader* (pp. 445–476). New York, NY: Oxford University Press.

Bhattacharya, R., Devinney, T.M., and Pillutla, M.M. (1998). A formal model of trust based on outcomes. *Academy of Management Review*, *23*(3), 459–472.

Campbell, D.J. and Campbell, K.M. (2009). Embracing change: Further examination of a 'capabilities and benevolence' beliefs model in a military sample. *Military Psychology*, *21*, 351–364.

Chiles, T.H. and McMackin, J.F. (1996). Integrating variable risk preferences, trust, and transaction cost economics. *Academy of Management Review*, *21*(1), 73–99.

Chu, P.C. and Spires, E.E. (2001). Does time constraint on users negate the efficacy of decision support systems? *Organizational Behavior and Human Decision Processes*, *85*, 226–249.

Colquitt, J.A., Scott, B.A., and LePine, J.A. (2007). Trust, trustworthiness, and trust propensity: A meta-analytic test of their unique relationships with risk taking and job performance. *Journal of Applied Psychology*, *92*(4), 909–927.

Dasgupta, P. (1988). Trust as a commodity. In D. Gambetta (ed.), *Trust: Making and Breaking Cooperative Relations* (pp. 49–72). New York, NY: Basil Blackwell.

Deluga, R. (1995). The relation between trust in the supervisor and subordinate organizational citizenship behavior. *Military Psychology*, *7*(1), 1.

Deutsch, M. (1958). Trust and suspicion. *Journal of Conflict Research*, *2*, 265–279.

Driskell, J.E. and Johnston, J.H. (1998). Stress exposure training. In J.A. Cannon-Bowers and E. Salas (eds), *Making Decisions under Stress: Implications for Individual and Team Training* (pp. 191–217). Washington, DC: American Psychology Association.

Easterbrook, J.A. (1959). The effect of emotion on cue utilization and the organization of behavior. *Psychological Review*, *66*, 187–201.

Edland, A. (1994). Time pressure and the application of decision rules: Choices and judgments among multiattribute alternatives. *Scandinavian Journal of Psychology*, *35*(3), 281–291.

French, J.R.P. Jr. and Raven, B. (1959). The bases of social power. In D. Cartwright (ed.), *Studies in Social Power* (pp. 150 – 167). Oxford, England: University of Michigan.

Frost, T., Stimpson, D.V., and Maughan, M.R. (1978). Some correlates of trust. *Journal of Psychology*, *99*, 103–108.

Gilbert, D. and Malone, P. (1995). The correspondence bias. *Psychological Bulletin*, *117*(1), 21–38.

Gurbuz, S. (2009). Some possible antecedents of military personnel organizational citizenship behavior. *Military Psychology*, *21*(2), 200–215.

Hayes, S.C., Bissett, R., Roget, N., Padilla, M., Kohlenberg, B.S., Fisher, G., Masuda, A., Pistoreleo, J., Rye, A.K., Berry, K., and Niccolls, R. (2004). The impact of acceptance and commitment training and multicultural training on the stigmatizing attitudes and professional burnout of substance abuse counselors. *Behavior Therapy*, *35*, 821–835.

Isenberg, D.J. (1981). Some effects of time-pressure on vertical structure and decision-making accuracy in small groups. *Organizational Behavior and Human Performance*, *27*(1), 119.

Kohn, H. (1954). Effects of variations of intensity of experimentally induced stress situations upon certain aspects of perception and performance. *Journal of Genetic Psychology*, *85*, 289–304.

Lewicki, R.J., McAllister, D.J., and Bies, R.J. (1998). Trust and distrust: New relationships and realities. *Academy of Management Review*, *23*(3), 438–458.

Maule, A., Hockey, G., and Bdzola, L. (2000). Effects of time-pressure on decision-making under uncertainty: Changes in affective states and information processing strategy. *Acta Psychologica, 104*(3), 283–301.

Mayer, R.C., Davis, J.H., and Schoorman, F.D. (1995). An integrative model of organizational trust. *The Academy of Management Review, 20*(3), 709–734.

McGrath, J., Arrow, H., and Berdahl, J. (2000). The study of groups: Past, present, and future. *Personality and Social Psychology Review (Lawrence Erlbaum Associates), 4*(1), 95–105.

McKinney, E.H. Jr., Barker, J.R., Davis, K.J., and Smith, D.R. (2005). How swift starting action teams get off the ground. *Management Communication Quarterly, 19*(2), 198–237.

McKinney, E.H. Jr., Barker, J.R., Smith, D.R., and Davis, K.J. (2004). The role of communication values in swift starting action teams: IT insights from flight crew experience. *Information and Management, 41*, 1043–1056.

Meyerson, D., Weick, K., and Kramer, R. (1996). Swift trust and temporary groups. *Trust in organizations: Frontiers of Theory and Research* (pp. 166–195). Thousand Oaks, CA: Sage Publications, Inc.

O'Donohue, W., Hayes, S.C., and Fisher, J.E. (2003). *Cognitive Behavior Therapy: Applying Empirically Supported Techniques in Your Practice.* John Wiley and Sons.

Ordóñez, L. and Benson, L. (1997). Decisions under time pressure: How time constraint affects risky decision making. *Organizational Behavior and Human Decision Processes, 71*(2), 121–140.

Oser, R.L., Gualtieri, J.W., Cannon-Bowers, J.A., and Salas, E. (1999). Training team problem-solving skills: An event-based approach. *Computers and Human Behavior, 15*, 441–462.

Payne, J.W., Bettman, J.R., and Johnson, E.J. (1995). An information processing perspective on choice. In J. Busemeyer, D.L. Medin and R. Hastie (eds), *Decision Making from a Cognitive Perspective* (Vol. 32, pp. 137–175). San Diego, CA: Academic Press.

Rempel, J.K., Ross, M., and Holmes, J.G. (2001). Trust and communicated attributions in close relationships. *Journal of Personality and Social Psychology, 81*(1), 57–64.

Rousseau, D.M., Sitkin, S.B., Burt, R.S., and Camerer, C. (1998). Not so different after all: A cross-discipline view of trust. *Academy of Management Review, 23*(3), 393–404.

Salas, E., Burke, C.S., and Cannon-Bowers, J.A. (2002). What we know about designing and delivering team training: Tips and guidelines. In K. Kraiger (ed.), *Creating, Implementing, and Managing Effective Training and Development: State-of-the-art Lessons for Practice* (pp. 234–259). San Francisco, CA: Jossey-Bass.

Salas, E. and Cannon-Bowers, J.A. (2000). The anatomy of team training. In S. Tobias and J.D. Fletcher (eds), *Training and Retraining: A Handbook for Business, Industry, Government, and the Military* (pp. 312–335). New York, NY: Macmillan Reference.

Smith-Jentsch, K.A., Salas, E., and Baker, D.P. (1996). Principles and strategies for team training. In W. Karwowski (ed.), *International Encyclopedia of Ergonomics and Human Factors.* New York, NY: Taylor and Francis.

Stern, P. and Fineberg, H. (1996). *Understanding Risk.* Washington, DC: National Academy Press.

Sundstrom, E., McIntyre, M., Halfhill, T., and Richards, H. (2000). Work groups: From the Hawthorne studies to work teams of the 1990s and beyond. *Group Dynamics: Theory, Research, and Practice*, *4*(1), 44–67.

Tajfel, H. (1982). Social psychology of intergroup relations. *Annual Review of Psychology*, *33*, 1–39.

Van der Kleij, R., Lijkwan, J.T. E., Rasker, P.C., and De Dreu, C.K. W. (2009). Effects of time pressure and communication environment on team processes and outcomes in dyadic planning. *International Journal of Human–Computer Studies*, *67*(5), 411–423.

Zakay, D. (1993). The impact of time perception processes on decision making under time stress. *Time Pressure and Stress in Human Judgment and Decision Making* (pp. 59–72). New York, NY: Plenum Press.

# Chapter 6
# Trust in Distributed Operations

Anna T. Cianciolo
*Command Performance Research, Inc., Champaign, IL, USA*

Karen M. Evans
*Rochester Institute of Technology, Rochester, NY, USA*

Arwen H. DeCostanza
*US Army Research Institute, Aberdeen Proving Ground, MD, USA*

Linda G. Pierce
*Federal Aviation Administration, Oklahoma City, OK, USA*

## Introduction

As information technology has become an ever more central component of military operations, increasing attention is being paid to understanding the impact of geographic distribution and networked communications on collective performance (e.g., Gorman, Cooke, and Winner, 2006; Neville, Fowlkes, Walwanis-Nelson, and Bergondy-Wilhelm, 2003; Ruddy, 2007; Stagl et al., 2007). Understandably, the typical foci of this attention are shared situation awareness and team agility, both of which are expected to improve as a result of networking people together and to relate directly to operational success (e.g., Alberts, Garstka, and Stein, 2000).

Trust is also becoming an area of increasing interest as researchers consider the issues surrounding network security as well as the interpersonal challenges associated with large, diverse teams addressing complex contingencies using technology-mediated communications (Blatt, 2004; Bowman, 2007; Cho and Swami, 2009; Chan, Ivanic, Rivera, and Bowman, 2009; Ekman and Uhr, 2008; Kemp, 2003). Trust is believed to facilitate distributed collective performance by promoting interpersonal influence and such cooperative activities as information sharing and operational support (e.g., Abrams, Cross, Lesser, and Levin, 2003; Gronn, 2002; Staples and Webster, 2008). However, the process by which trust functions in distributed military operations is not well understood and is rarely investigated empirically.

We are currently conducting research funded by the US Army Research Institute to examine interpersonal trust in network-enabled command and control (C2) teams at the brigade and division level. Of particular interest is the way trust functions in such teams as they conduct joint, interagency, intergovernmental, and multinational (JIIM) operations. The ultimate goal of our work is to aid future C2 concepts experimentation in accounting for the impact of structural and procedural changes on trust and collective performance, and to identify how trust develops in network-enabled C2 teams. Our work supports a larger initiative aimed at enhancing collaboration and decision making in complex, network-enabled C2 environments (ATO R.ARL.2009.05/Tactical Human Integration of Networked Knowledge; THINK ATO). Toward this end, we seek to understand what trust 'looks like' in a distributed team's communications and patterns of information exchange – data that are automatically captured in C2 experiments. Over the course of our four-year study, we will construct a conceptual model of trust and investigate the causal links between environmental conditions and the affect and behavior that reflect trust. We will apply our findings to establish measures of trust and to develop approaches for fostering calibrated trust within distributed teams.

In this chapter we present the initial findings of our research. We propose a conceptual model of trust in distributed military operations that is based on a current, extensive literature review and qualitative research. We describe the implications of our model for empirical investigation and highlight the ways in which experiments conducted using our model may generalize across forms of distributed military operation.

## An Operational Definition of Trust

To operationally define trust, we begin by adopting Mayer, Davis, and Schoorman's (1995) conception, which is widely used in the trust literature. Mayer et al. defined trust as the willingness of an individual (or *trustor*) to be vulnerable to the actions of an object of trust (or *trustee*) regardless of external control or visibility on those actions. Despite fairly uniform notions of what trust is, there is marked diversity in how trust is operationally defined. For instance, there is disagreement over whether trust is fundamentally an affective state of the trustor (measurable via survey instruments) or a critical determinant of behavior (captured via self-report or observation of behavior). Researchers have advised against behavioral measurement of trust because behavior can be sensitive to many factors besides those of interest (e.g., Mayer et al., 1995; Rousseau, Sitkin, Burt, and Camerer, 1998). However, the imperfect relation between attitude and behavior raises questions about the relevance of investigating affect alone, especially when accounting for the impact of trust on performance is a research objective.

Extending previous research, we propose that willingness to be vulnerable is a process that can be defined and reliably measured as *both* affect and behavior. That is, trust requires feeling that an object is trustworthy *and* acting as though that object is trustworthy, even when there is risk involved in doing so (i.e., there is no way to control or oversee the trustee's behavior). This proposition requires explicit specification of the context in which trust is measured such that (1) the number of variables moderating the link between affect and behavior can be minimized and modeled statistically; and (2) the influence of environmental conditions on the trust process can be articulated.

In our research, we operationally define trust-related attitude as a set of expectations about a trustee's behavior in a given situation (i.e., a mission- or task-specific judgment of trustworthiness). This judgment may be measured using self-report methods in which the trustor, after being presented with the characteristics of a trustee, is asked to estimate the likelihood that the trustee will perform according to specific positive expectations (e.g., 'It is 90 percent likely that the intelligence officer will contribute favorably to mission success.'). We define trust-related behavior as the degree to which a trustor attempts to mitigate risk in the face of perceived vulnerability. Risk-mitigation behaviors may be captured in the nature and relative patterns of communications among more and less trusted members of the team (e.g., frequency of 'one-way conversations;' see, e.g., Barron, 2000; Xiao, Seagull, Mackenzie, Ziegert, and Klein, 2003). Given that a trustor may make a range of predictions about how others will behave and given that there are diverse behaviors that can be construed as risk management, a critical part of our research is identifying the precise expectations that are associated with trust-related behavior and determining the risk-management behaviors that have meaningful implications for collective performance.

## Why and How to Study Trust in Distributed Operations

On the surface, it may seem obvious why trust would be important in distributed military operations. Military engagements are high risk and often involve a great deal of uncertainty, thus creating a strong sense of vulnerability; participants must have confidence that the others involved will enhance (or at least not hurt) their chances of survival. A closer look at the operating environment of distributed military teams, however, offers a more actionable account of how judgments of trustworthiness form and influence behavior and how the trust process should be studied. It reveals the conditions that must be present in an experimental situation for positive expectations to influence behavior, as well as the factors that must be manipulated to affect their formation.

The most salient characteristic of distributed military teams is the geographic dispersion of their members. Dispersion challenges the development of positive expectations because there are fewer occasions for face-to-face contact, and the prevailing communication methods do not convey the same information or do so only over longer periods of time (Hung, Dennis, and Robert, 2004; Kasper-Fuehrer and Ashkanasy, 2001; Riegelsberger, Sasse, and McCarthy, 2003). Technology-mediated communications also introduce the possibility that technical problems or usability shortfalls will be attributed to people, increasing the likelihood of false impressions of low trustworthiness. To understand trust in distributed teams, therefore, it must be determined who/what is being trusted and how to overcome the challenges posed by technology-mediated communications. In empirical investigations, it is critical to ensure that the relative frequency and function of face-to-face versus other methods of communication is preserved. Moreover, the information technology used should feature capabilities that research participants normally use to conduct collaborative work (and, by extension, to convey their trustworthiness). Finally, each mode of communication available should present a level of trustworthiness that is consistent with operating conditions. In network-enabled teams, information technology itself becomes an object of trust (Blatt, 2004), which can affect the signature of risk-management behavior independently of positive expectations for people (e.g., how and when team members will use a network versus other means to share information). Measures of behavior must be sensitive to the different signatures of positive expectations as they relate to people versus the networked collective.

Perhaps more important to trust formation than technology-mediated communications, assembling distributed teams makes it possible to take on more complex, ill-defined problems using groups of people who are rapidly convened, diverse, and structured horizontally such that decision-making power is distributed, rather than held by a single individual at the top of a hierarchy (Gronn, 2002). Distributed decision making removes external controls, such as a unified reporting structure that requires people to work interdependently to achieve personal success. This 'functional independence' makes interpersonal relations a primary driver of cohesive, coordinated action (Cianciolo, LaVoie, Folz, and Pierce, 2009; Gronn, 2002; Staples and Webster, 2008). Team diversity and ill-defined problems make it more difficult for members to judge each other's contributions because they lack the requisite knowledge to process the available information (e.g., a soldier may not be able to judge the expertise of an airman; the lack of a single correct approach introduces subjectivity to assessing others' contributions). In the absence of established working relationships – a common condition in distributed military teams – team members may arrive at judgments by using assumptions and stereotypes based on each others' social identity (e.g., whether their unit differs from the trustor's; Adams, Waldherr, Sartori and Thomson, 2007). For these reasons, it is critical to ensure that the problems being solved collectively in trust experiments appropriately represent operational problem complexity, team diversity, and level of functional independence among research participants. Failure to represent the operational environment in this way would produce approaches to measuring and developing trust that have minimal impact on actual distributed military teams.

Distributed C2 teams conducting JIIM operations are very large-scale, and they are characterized by blended structure (hierarchical and horizontal) as well as blended communications (face-to-face and technology-mediated). The hybrid nature of such teams makes it possible to define distinct units of analysis that a model of trust should account for. By unit of analysis, we do not mean the scale of the object studied (e.g., dyads, small teams, entire network-enabled C2 team). Rather, we are referring to subsets of the entire network-enabled team that have implications for how the trust process functions. That is, these subsets differ in terms of level of geographic distribution, amount of uncertainty associated with collective work, diversity, and the other team characteristics described above. Example units of analysis include a co-located command group, a distributed warfighting function (e.g., brigade and division intelligence cells), and functionally independent interagency planning teams, among others. Specifying units of analysis forms a taxonomy of 'ecological niches' in which distributed operations can be conducted, each of which represents broad types of distributed team or operation to which the trust process should generalize. Below we present our conceptual model of trust, which generalizes across units of analysis. Next we describe a proposed set of units of analysis for enabling generalizable trust research design.

## A High-Level Conceptual Model of Trust

The findings presented in this chapter reflect a combination of extensive literature review and qualitative analysis in the domain of interest (i.e., brigade and higher C2 during JIIM operations; see also Evans, Cianciolo, Hunter, and Pierce, 2010). Our literature review covered a wide range of theory papers and empirical investigations, including explorations of trust in commercial organizations, work teams, close personal relationships, leader–follower dyads, and small military units. We also reviewed literature that examined trust in automation, in information technology, and in networked teams, organizations, and communities. The purpose of the literature review was to identify emergent, high-level patterns among trust-related affect and behavior as influenced by situational context. We conducted qualitative analysis to explore in depth the particular context to which we would be applying the results of our literature review. We first conducted one-on-one interviews with eight active duty or recently retired Army personnel who had served either as a commander or key staff member of a battalion- or brigade-level command group. We also observed a division-level C2 exercise conducted with students at the Command and General Staff College located at Fort Leavenworth, Kansas and two division-level, multinational C2 exercises (Omni Fusion 2008 and 2009), conducted at the Fort Leavenworth Battle Command Battle Lab. Finally, we reviewed the reports of trained analysts who attended the Omni Fusion exercises.

Figure 6.1 depicts our overarching conceptual model of trust in distributed operations. Consistent with extant trust models (e.g., Adams and Webb, 2003; Colquitt, Scott, and LePine, 2007; Mayer et al., 1995), the left side of the figure posits that some combination of trustee behaviors and trustor characteristics influences judgments of trustworthiness. Trustee behaviors are concrete instantiations of latent trustworthiness dimensions and are specific to the military operation being conducted. Trustor characteristics are general personal characteristics associated with positive expectations of others (e.g., propensity to trust; Rotter, 1980). This component of the model accounts for both situational and personal influences on a trustor's expectations of a trustee, which vary in strength depending on the situation (Colquitt et al., 2007; Robert, Dennis, and Hung, 2009). On the right side of the figure, expectations are shown affecting a trustor's behavior, which is a concrete indicator of the trustor's level of willingness to accept risk (e.g., attempts to gain control of a situation reflect a need to protect oneself against perceived vulnerability). The relationship between affect and behavior is moderated both by situational conditions (e.g., risk) and personal characteristics of the trustor associated with risk taking (e.g., risk aversion, need for control).

Importantly, the unit of analysis, shown in the figure below as encompassing the trust process, influences several aspects of the conceptual model and allows explicit specification of how the trust process will differ for various types of distributed military team. For instance, we posit that although the latent dimensions of trustworthiness shown in Figure 6.1 remain the same across units of analysis, their relative importance and behavioral manifestation will differ, thus affecting the cues of trustworthiness that trustors use to form expectations. The unit of analysis also influences the risk-management options available to trustors because each is associated with a different mix of communication methods. Finally, by representing different levels of uncertainty, functional independence, and other situational conditions, the unit of analysis drives which moderators influence the relation between trust-related attitude and behavior. Specifying units of analysis enables the simultaneous study of a variety of operational applications while managing complexity by explicitly modeling both C2 teams and the conditions in which they work.

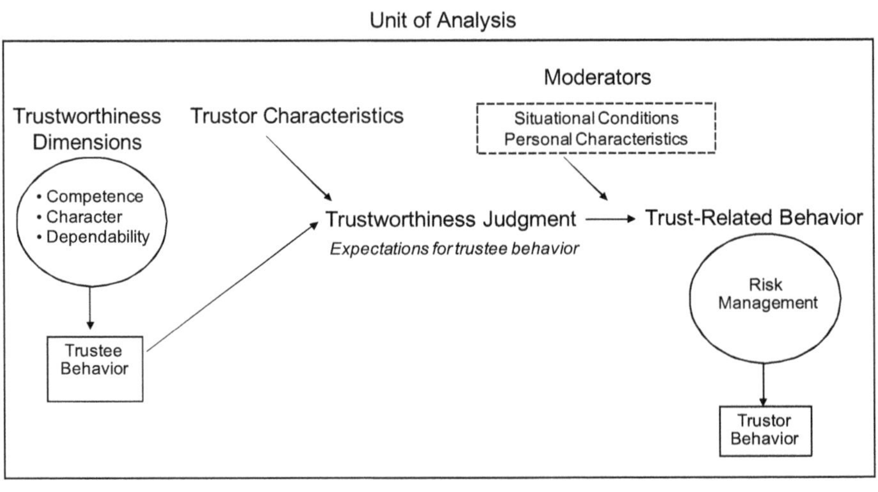


**Figure 6.1 High-level conceptual model of trust**

*Determinants of Trustworthiness Judgments*

Recall that distributed operations introduce the possibility of both human and technological trustees (Blatt, 2004). Our model posits that when judging objects of trust, trustors rely on cues that convey that the trustee is willing and able to do consistently what is needed. These cues, respectively, represent the latent trustworthiness dimensions of character, competence, and dependability. These dimensions appear universal across relationship types, including coworkers (e.g., Colquitt et al., 2007; Gabarro, 1978; Mayer et al., 1995; McAllister, 1995), distributed or co-located project members (Jarvenpaa, Knoll, and Leidner, 1998; Serva, Fuller, and Mayer, 2005), and military teams (Adams and Webb, 2003; Ruark, Orvis, Horn, and Langkamer, 2009; Sweeney, 2007). These dimensions also seem to apply to networked teams (i.e., group competence, group character, and group dependability; Adams and Webb, 2003; Ardichvili, Page, and Wentling, 2003; Jarvenpaa et al., 1998; Usoro, Sharratt, Tsui, and Shekhar, 2007) and to technology, where character reflects the alignment of the designer's intent with operator's use of the system, and competence and dependability are reflected in system performance (reliability, predictability, capability, and usability; adopted from Lee and See, 2004).

*Competence* Military research on trust has mostly focused on combat teams and has found that although competence is essential to judging a commander's trustworthiness (Sweeney, 2007), it may be less important than character and dependability when assessing peers and subordinates (Adams and Webb, 2003; Ruark et al., 2009). In contrast, our qualitative research revealed that competence is essential to the trustworthiness of all actors in battalion and higher C2 (Evans et al., 2010). The importance of a networked collective's competence has been highlighted in research on online learning communities, where community members reported more positive expectations and engaged in fewer risk-mitigation behaviors when they perceived high competence in the group (e.g., Ardichvili et al., 2003; Usoro et al., 2007). Similarly, we observed that participants in C2 exercises were reluctant to use the network's human-entered data when they perceived little competence in their fellow network operators.

Cues of individual competence in C2 may be gathered from information-sharing behaviors and general problem-solving and planning behaviors, as well as from indirect sources such as group membership (e.g., unit; Adams et al., 2007), depending on how long the team has worked together. Example behavioral cues include providing actionable information, anticipating information requirements, and identifying the core issues of problems. Although judging the group's competence might be accomplished by aggregating the competence of individual group members, our research uncovered cues unique to group-level competence, including group-level information-sharing behaviors (e.g., frequency of errors on the network), actual or perceived capability of the group, and stereotypes about the competence of the groups that constitute the collective.

*Character* Positive trustworthiness judgments of an individual trustee may reflect belief that the trustee will be kind to the trustor specifically (benevolence, loyalty, lack of opportunism) or that the trustee's general principles promote prosocial behavior (integrity, honesty, 'team player' attitude; Colquitt et al., 2007; Gabarro, 1978; Mayer et al., 1995; Serva et al., 2005). Echoing claims about the importance of character to trust in small military units (Adams and Webb, 2003; Ruark et al., 2009; Sweeney, 2007), our investigation has found that character is important for judging individual trustworthiness in network-enabled C2. However, our interviewees reported believing that character strength is fairly homogeneous across command group members, and that the character of network operators need not be questioned given that personnel with access to C2 systems are military and therefore share similar values.

A salient cue of both individual and collective character appears to be group membership. Although our interviewees said that fellow military personnel (especially Army personnel) are assumed to possess strong character, they reported that JIIM partners are often initially *dis*trusted, given the likelihood that they possess different values and goals, and may even compete for resources (Evans et al., 2010). Behavioral cues of individual character include helping, information sharing, and actions that reflect the trustee's sense of responsibility and accountability to the team and its mission. Collective character may be inferred from group diversity, given the importance of social identity in judging the trustworthiness of unfamiliar collaboration partners (Adams et al., 2007).

*Dependability* Although several researchers characterize dependability as a component of character or even competence, dependability can be dissociated from intention and accuracy (e.g., a highly competent partner may provide information in an irregular and untimely fashion due to information overload rather than disinterest in the trustor's welfare). This distinct ability to count on and predict a trustee's behavior seems to be especially important for trust in combat teams (Adams and Webb, 2003; Ruark et al., 2009), and our research has found that individual dependability is also essential to information exchange in network-enabled C2. Group dependability is important because it has implications for whether network operators can assume (in the absence of control) consistency in the information they receive and in the responsiveness to information they contribute. Even when members are competent and well meaning, the quality of the information they share could still be compromised by difficulties operating in a shared virtual space. Members of the collective must therefore trust that their peers can communicate effectively and that the information technology that unites them transmits data faithfully.

Our research suggests that behavioral cues of individual dependability include providing timely information and updates, being available/accessible when needed, and following through on commitments. For unfamiliar collaboration partners, group membership or reputation (from a trusted third party) may be used to infer dependability. Behavioral cues of group dependability include adoption of standard operating procedures for information management and the currency of information found on the network. Importantly, group dependability cannot be inferred from a simple average of individual behavioral cues; the consistency of the collective can be distorted by the actions of a single member. Group dependability is evidenced in the tendency of all members of the collective to contribute and respond to information on the network in predictable ways, and this was a sizable hurdle in the military simulations we observed in our research.

*Personal characteristics* In addition to characteristics of the trustee, characteristics of the trustor can influence judgments of trustworthiness, particularly in new relationships where there is limited exposure to behavioral cues (Colquitt et al., 2007; Kiffin-Petersen and Cordery, 2003; Mayer et al., 1995). In distributed communications, collaborators receive less information about each other (Hung et al., 2004; Riegelsberger et al., 2003), which increases the likelihood that personal characteristics will drive expectations about others. The most frequently studied of these characteristics is the trustor's propensity to trust others, which refers to the likelihood that the trustor will make favorable projections in the absence of any other information (Rotter, 1980). Some researchers have also suggested that personality characteristics such as high agreeableness and low neuroticism (Dirks and Skarlicki, 2004) might influence trustworthiness judgments. The influence of personal characteristics in judging the trustworthiness of the networked collective is an area in need of further research; however, there is substantial evidence that the use of technology is sensitive to one's propensity to trust technology (Dzindolet, Dawe, Beck, and Pierce, 2001; Lee and See, 2004; Merritt and Ilgen, 2008). Judgments of the trustworthiness of information technology in distributed teams might reflect this trustor characteristic.

### *Trust-related Behaviors*

Expectations regarding the outcome associated with engaging a trustee influence perceived vulnerability. Risk management is a set of behaviors that directly reflects this perception (as opposed to other factors, such as perceptions of likeability and understanding of roles and responsibilities). Behaviors associated with having positive expectations about a person include open and honest communication and knowledge sharing (Adams, Bryant, and Webb, 2001; Chowdhury, 2005; Staples and Webster, 2008; Zand, 1972) and collaboration (Adams et al., 2007; Jarvenpaa and Shaw, 1998). Positive expectations for the networked collective promote knowledge-sharing behaviors using information technology, such as communicating freely on the group network and readily accepting information found on the network (Abrams et al., 2003; Ardichvili et al., 2003; Usoro et al., 2007). Our qualitative research indicated that people behave similarly towards trustworthy people and collectives; trustors are more likely to accept information from these sources, less likely to double-check exchanges, and more likely to make contributions.

When the outcomes of interacting with a trustee are expected to be negative, the trustor reduces his perceived vulnerability by taking action to mitigate risk. Risk is managed by prioritizing mitigating actions according to greatest degree of perceived vulnerability. Distrust in people has been associated with monitoring the trustee's behavior, giving the trustee extra time or duplicating his or her efforts, creating formal rules, requests, and contracts, and limiting the trustee's responsibilities by assigning fewer or lower-stakes tasks (McAllister, 1995; Serva et al., 2005). Our qualitative research further indicated that C2 team members often mitigate risk by reducing their interactions with the trustee (e.g., becoming less accessible), changing the nature of interactions with the trustee (e.g., offering guidance rather than seeking guidance or information), and evaluating the trustee's work (e.g., closely double-checking or monitoring). Risk-mitigation behaviors in response to an untrusted collective appear to be similar in nature.

The nature of risk-mitigation activities may depend on the source of negative expectations. Specifically, trustees who are deemed well intentioned and dependable, but not competent, are likely to be instructed so that their competence can improve. In contrast, belief that the trustee is lazy or self-serving cannot be remedied as easily, so the trustor is likely to avoid this person where possible, especially in social situations. Risk-mitigation behaviors are also sensitive to hierarchical authority, since trustors with authority over the trustee have a greater range of options available to them (e.g., obviation, reassignment; Vogelaar and Kramer, 2000). Overall, communication patterns associated with less trust would be marked by reduced frequency of interaction with the trustee and increased exchanges with others to verify information.

### *Situational Factors*

Our model posits that the link between judgments of trustworthiness and risk-management behaviors is moderated by situational conditions and personality variables. These factors may promote trust-related behavior in the absence of trust-related attitudes or, conversely, prompt risk-mitigation activities even when judged trustworthiness is high. Because full models of trust have never been empirically tested, the exact function of moderator variables and their relative importance remains unknown. Our research suggests a few high-payoff moderators that may apply across both objects of trust and await further investigation.

Risk is commonly identified as a key moderator (e.g., Adams and Webb, 2003; Hung et al., 2004; Mayer et al., 1995), and drives the perceived cost associated with action. Perceived cost may increase risk-mitigation activities regardless of trust-related attitudes. For example, a trustor may double-check the work of a highly trusted individual or group when there is a lot at stake and when errors, however unlikely, could be very harmful. However, risk may reduce risk-mitigation activities whose outcomes are more uncertain or difficult to control. For example, one of our interviewees reported that training missions were better opportunities (than deployment) to develop less trustworthy personnel by providing direction and mentorship. In this case, more frequent interactions with less trustworthy personnel would be more common in lower-risk situations. This same pattern may emerge with less trustworthy networked collectives as they use lower-risk training conditions to test procedures and develop operating standards.

Some situational conditions influence the relation between affect and behavior by directly constraining opportunity to mitigate risk. If there is no other choice or there is insufficient time, trustors may forgo risk-mitigation actions, such as reassigning the task to a more trusted person or doing the task alone, simply because these actions are not possible (Lee and See, 2004; Mayer et al., 1995). Research has found that the degree of task interdependence or the presence of formal rules or protocols guiding information exchange may structure interactions in a way that weakens the statistical relationship between judgment and behavior (Butler, 1999; Jarvenpaa, Shaw, and Staples, 2004; Staples and Webster, 2008). Similarly, the adoption of standing operating procedures (SOPs) that dictate the preferred information flow in network-enabled C2 teams could limit how closely communication patterns mirror trustworthiness judgments. When characterizing situational conditions, it is important to distinguish between actual and perceived constraints on alternatives. For instance, in distributed C2, especially when decision making is distributed, team members may be ordered to execute a shared mission, but can nevertheless act independently with impunity due to a lack of accountability. In our observations of simulated C2, we noted weaker relations between trust-related attitudes and reported information sharing when perceived interdependence was high.

Some researchers have theorized that personality traits of the trustor also influence the relationship between trustworthiness judgments and behavior. For example, cultural or personal values that promote trusting behaviors might encourage trustors to engage in such behaviors even when positive expectations are weak or non-existent (Whitener, Brodt, Korsgaard, and Werner, 1998). Traits such as risk aversion could lead to individuals to place a higher value on avoiding costs associated with mistakes and thereby promote risk-mitigation activities regardless of positive expectations.

## Units of Analysis for Modeling Trust

As described previously, the unit of analysis chosen for study influences several aspects of the conceptual model and has implications for research design, statistical modeling, and generalizability. Most trust experiments involve only one, unspecified unit of analysis, which limits the generalizability of the findings and has produced a wealth of inconsistent results. We propose three units of analysis that will enable concise specification of model parameters for our research and clear indication of the other types of military team to which our findings should apply. To define units of analysis, we identified a set of collective types that simultaneously (1) represented functional groupings naturally occurring in network-enabled C2 teams conducting JIIM operations; and (2) varied along key situational conditions that relate to trust processes (see Table 6.1 below). The collectives we identified are uniform in size and relatively small (approximately five personnel), which enables feasible experimentation that is likely to generalize to the application environment.

We defined *nuclear* collectives as those involving strictly Army personnel, with membership limited to the commander and key staff members with whom the commander coordinates most closely (e.g., the operations and intelligence cell chiefs). The nuclear collective specifically excludes staff duty positions typically kept at the outskirts of communications and information sharing due to role ambiguity or the tendency to 'stovepipe' staff activities (e.g., personnel and logistics staff officers; Cianciolo and Sanders, 2006). Because the roles among duty positions selected for nuclear collectives are well understood by the personnel involved, the likelihood is reduced that factors other than trust, such as role knowledge, will cloud interpretation of communications and information-sharing patterns. In network-enabled C2 teams, nuclear collectives tend to communicate primarily face-to-face, are relatively homogeneous, and are characterized by well-established working relationships, which minimizes uncertainty with regard to collective work and consequently enhances confidence in judgments of trustworthiness. Within nuclear collectives, trustworthiness judgments are more likely to be determined by trustee characteristics (than trustor personality), and character may play as strong a role in positive expectations as competence and dependability (Colquitt et al., 2007; Jarvenpaa et al., 1998). Nuclear collectives most closely resemble small combat units in terms of their composition, although the nature of risk across the two types of team may differ (direct versus indirect threat of bodily harm).

**Table 6.1 Variation as a function of collective type**

| Collective type | Relationship history | Distribution | Independence (functional) | Confidence in judgment | Uncertainty |
|---|---|---|---|---|---|
| Nuclear | Long | Co-located | High | High | Low |
| Echelon | Long/ moderate | Distributed, interoperable | Moderate | Moderate | Moderate |
| Diverse | Short | Distributed, non-interoperable | Low | Low | High |

*Echelon* collectives are also exclusively comprised of Army personnel but span across echelons within a particular warfighting function (e.g., intelligence, maneuver, fires, etc.). The primary difference between echelon and nuclear collectives is geographic distribution; whereas members of nuclear collectives are typically co-located, members of echelon collectives typically are not, although they share expertise, a common operating picture, and interoperable communications devices. Clearly, in this type of collective, information technology becomes more important as an object of trust and as a medium for assessing behavior than in nuclear collectives. Uncertainty with regard to collective work may increase (and confidence in trustworthiness judgments decease) due to time delays, geographic distance, and technical errors introduced by technology-mediated communications. Echelon collectives are characterized by relatively well-established working relationships, although the personal or affective bonds between members may be weaker than those in nuclear collectives. In this way, echelon collectives represent other single-service distributed military teams studied in the literature. When defining an echelon collective and determining representativeness, it is critical to account for whether the collective contains a direct chain of command (i.e., superior and subordinate commanders vs. staff officers of mixed rank). The presence of a direct chain of command can influence the level of perceived interdependence, which, as stated earlier, moderates the link between trust-related attitudes and information sharing (Staples and Webster, 2008).

*Diverse* collectives involve Army personnel from different nations, including the US (e.g., an American division commander, British brigade commanders, and possibly liaison officers) and/or Army personnel and members of civilian government agencies and/or personnel from other military services (e.g., a campaign planning team or a division fires cell that includes Air Force liaisons). Like echelon collectives, diverse collectives involve geographic distribution and technology-mediated communication. In contrast to echelon collectives, diverse collectives are characterized by (1) non-interoperable C2 systems; (2) cultural diversity; and (3) lack of well-established interpersonal relationships, which further increase uncertainty and decrease confidence in trustworthiness judgments. Diverse teams are also characterized by high functional independence, due to the lack of a unified chain of command (i.e., separation of powers). It may be expected that the link between trust-related affect and behavior may be stronger in this type of collective than in nuclear or echelon collectives with a direct chain of command. In addition, the importance of trustor characteristics (i.e., social identity) may increase in importance for determining positive expectations about other members of the collective (Colquitt et al., 2007; Kiffin-Petersen and Cordery, 2003; Mayer et al., 1995). Findings based on this type of collective in network-enabled C2 teams may generalize to distributed multinational operations that are smaller in scale (e.g., at lower echelons), to other interagency collectives such as disaster relief teams, or to smaller scale multi-service teams.

## Conclusion

The present chapter summarizes our approach to understanding trust in distributed military operations. Although our focus is on network-enabled C2 teams conducting JIIM operations, the approach we have adopted allows us to specify the other types of distributed military teams to which our findings should generalize. Driven by the need to account for the impact of trust on behavior, we have developed a conceptual model of trust and associated units of analysis that differ substantially from the extant body of trust research, which has focused alternately on trust-related affect or behavior, but rarely on both simultaneously. In our current research, this approach will enable us to account for the impact of changes to C2 structure and process on interpersonal trust. With modification to the specific behavioral indicators of trustworthiness and risk management our approach should enable the investigation of this human dimension across a wide variety of distributed military teams.

## References

Abrams, L. C., Cross, R., Lesser, E., and Levin, D. Z. (2003). Nurturing interpersonal trust in knowledge-sharing networks. *Academy of Management Review*, *17*(4), 64–77.

Adams, B. D., Waldherr, S., Sartori, J., and Thomson, M. (2007). *Swift trust in distributed ad hoc teams* (DRDC Toronto No. CR-2007–139). Toronto, Canada: Department of National Defence, Defence Research and Development Canada.

Adams, B. D. and Webb R. D. G. (2003). *Trust development in small teams* (DRDC Toronto No. CR-2003-016). Toronto, Canada: Department of National Defence, Defence Research and Development Canada.

Adams, B. D., Bryant, D. J., and Webb, R. D. G. (2001). *Trust in teams: Literature review* (Rep. No. CR-2001–042). Toronto, Canada: Department of National Defence, Defence Research and Development Canada.

Alberts, D. S., Garstka, J. J., and Stein, F. P. (2000). *Network centric warfare: Developing and leveraging information superiority* (2nd rev. edn). Washington, DC: Department of Defense Command and Control Research Program Publications Series.

Ardichvili, A., Page, V., and Wentling, T. (2003). Motivation and barriers to participation in virtual knowledge-sharing communities of practice. *Journal of Knowledge Management*, *7*(1), 64–77.

Barron, B. (2000). Achieving coordination in collaborative problem-solving groups. *The Journal of the Learning Sciences*, *9*(4), 403–436.

Blatt, N. (2004). Operational trust: A new look at the human requirement in network centric warfare. *Paper presented at the 9th International Command and Control Research and Technology Symposium on Coalition Transformation.* Copenhagen, Denmark.

Bowman, A. K. (2007). Measuring team collaboration in a distributed coalition network. *Twelfth International Command and Control Research and Technology Symposium*. Newport, RI.

Butler, J. K., Jr. (1999). Trust expectations, information sharing, climate of trust, and negotiation effectiveness and efficiency. *Group and Organization Management, 24*, 217–238.

Chan, K. S., Ivanic, N., Rivera, B., and Bowman, E. K. (2009). A general framework of human trust in networks. *Proceedings of the 14th International Command and Control Research and Technology Symposium*. Santa Monica, CA.

Cho, J. and Swami, A. (2009). Towards trust-based cognitive networks: A survey of trust management for mobile ad hoc networks. *Proceedings of the 14th International Command and Control Research and Technology Symposium.* Santa Monica, CA.

Chowdhury, S. (2005). The role of affect- and cognition-based trust in complex knowledge sharing. *Journal of Managerial Issues, 17*(3), 310–326.

Cianciolo, A. T., LaVoie, N., Foltz, P., and Pierce, L. G. (2009). *Augmented performance environment for enhancing interagency coordination in stability, security, transition, and reconstruction (SSTR) operations* (Tech. Rep. 1246). Arlington, VA: US Army Research Institute for the Behavioral and Social Sciences.

Cianciolo, A. T. and Sanders, W. R. (2006). *Wargaming effectiveness: Its conceptualization and assessment* (Tech. Rep. 1178). Arlington, VA: US Army Research Institute for the Behavioral and Social Sciences.

Colquitt, J. A., Scott, B. A., and LePine, J. A. (2007). Trust, trustworthiness, and trust propensity: A meta-analytic test of their unique relationships with risk taking and job performance. *Journal of Applied Psychology, 92*(4), 909–927.

Dirks, K. T. and Skarlicki, D. (2004). Trust in leaders: Existing research and emerging issues. In R. Kramer and K. Cook (eds), *Trust and distrust in organizations: Dilemmas and approaches* (pp. 21–40). New York: Russell Sage Foundation.

Dzindolet, M. T., Dawe, L. A., Beck, H. P., and Pierce, L. G. (2001). *A framework of automation use* (Report No. ARL-TR-2412). Aberdeen Proving Ground, MD: Army Research Laboratory.

Ekman, O. and Uhr, C. (2008). Crisis specific social networks: The interplay between organisational legitimacy and personal trust. *Proceedings of the 13th International Command and Control Research and Technology Symposium.* Bellevue, WA.

Evans K. M., Cianciolo, A. T., Hunter, A. E., and Pierce, L. G. (2010). Modeling interpersonal trust in distributed command and control teams. *Proceedings of the 15th International Command and Control Research and Technology Symposium*. Santa Monica, CA.

Gabarro, J. J. (1978). The development of trust, influence, and expectations. In A. Athos and J. J. Gabarro (eds), *Interpersonal behavior: Communication and understanding in relationships* (pp. 290–303). Englewood Cliffs, NJ: Prentice-Hall.

Gorman, J. C., Cooke, N. J., and Winner, J. L. (2006). Measuring team situation awareness in decentralized command and control environments. *Ergonomics*, *49*(12/13), 1312–1325.

Gronn, P. (2002). Distributed leadership as a unit of analysis. *The Leadership Quarterly*, *13*, 423–451.

Hung, Y-T, C., Dennis, A. R., and Robert, L. (2004). Trust in virtual teams: Towards an integrative model of trust formation. *Proceedings of the 37th Hawaii International Conference on System Sciences* (pp. 1–11). New York: Institute of Electrical and Electronics Engineers.

Jarvenpaa, S. L. and Shaw, T. R. (1998). Global virtual teams: Integrating models of trust. In P. Sieber and J. Griese (eds), *Organizational virtualness: Proceedings of the VoNet Workshop* (pp. 35–51). Switzerland: Simowa Verlag Bern.

Jarvenpaa, S. L., Knoll, K., and Leidner, D. E. (1998). Anybody out there? Antecedents of trust in global virtual teams. *Journal of Management Information Systems*, *14*(4), 29–64.

Jarvenpaa, S. L., Shaw, T. R., and Staples, D.S. (2004). Toward contextualized theories of trust: the role of trust in global virtual teams. *Information Systems Research*, *15*(3), 250–267.

Kasper-Fuehrer, E. C. and Ashkanasy, N. M. (2001). Communicating trust-worthiness and building trust in interorganizational virtual organizations. *Journal of Management 27*, 235–254.

Kemp, C. R. (2003). *Trust – The key to leadership in network centric environments*. Carlisle, PA: US Army War College Strategy Research Project.

Kiffin-Petersen, S. A. and Cordery, J. L. (2003). Trust, individualism, and job characteristics as predictors of employee preference for teamwork. *International Journal of Human Resource Management*, *14*(1), 93–116.

Lee, J. D. and See, K. A. (2004). Trust in automation: Designing for appropriate reliance. *Human Factors*, *46*(1), 50–80.

Mayer, R. C., Davis, J. H., and Schoorman, F. D. (1995). An integrative model of organizational trust. *Academy of Management Review*, *20*, 709–734.

McAllister, D. (1995). Affect- and cognition-based trust as foundations for interpersonal cooperation in organizations. *Academy of Management Journal*, *38*, 24–59.

Merritt, S. M. and Ilgen, D. R. (2008). Not all trust is created equal (individual differences in trust). *Human Factors*, *50*(2), 194–210.

Neville, K., Fowlkes, J. E., Walwanis-Nelson, M. M., and Bergondy-Wilhelm, M. L. (2003). A cognitive task analysis of coordination in a distributed tactical team: Implications for expertise acquisition. *Proceedings of the 47th Annual Meeting of the Human Factors and Ergonomics Society* (pp. 359–363). Santa Monica, CA: Human Factors and Ergonomics Society.

Riegelsberger, J., Sasse, M. A., and McCarthy, J. D. (2003). The researcher's dilemma: Evaluating trust in computer-mediated communication. *International Journal of Human-Computer Studies, 58*, 759–781.

Robert, L. P., Jr., Dennis, A. R., and Hung, Y-T. C. (2009). Individual swift trust and knowledge-based trust in face-to-face and virtual team members. *Journal of Management Information Systems, 26*(2), 241–279.

Rotter, J. B. (1980). Interpersonal trust, trustworthiness, and gullibility. *American Psychologist, 35*(1), 1–7.

Rousseau, D. M., Sitkin, S. B., Burt, S. B., and Camerer, C. (1998). Not so different after all: Across-discipline view of trust. *Academy of Management Review, 23*(3), 393–404.

Ruark, G. A., Orvis, K. L., Horn, Z., and Langkamer, K. L. (2009). Trust as defined by U.S. Army Soldiers. *Proceedings of the 53rd Annual Meeting of the Human Factors and Ergonomics Society* (pp. 1924–1928). Santa Monica, CA: Human Factors and Ergonomics Society.

Ruddy, M. (2007). ELICIT – The experimental laboratory for investigating collaboration, information-sharing, and trust. *Proceedings of the Command and Control Research and Technology Symposium*. Newport, RI.

Serva, M. A., Fuller, M. A., and Mayer, R. C. (2005). The reciprocal nature of trust: A longitudinal study of interacting teams. *Journal of Organizational Behavior, 26*(6), 625–648.

Stagl, K. C., Salas, E., Rosen, M. A., Priest, H. A., Burke, S., Goodwin, G. F., and Johnston, J. H. (2007). Distributed team performance: A multi-level review of distribution, demography, and decision making. In F. Dansereau and F. J. Yammarino (eds), *Multi level issues in organizations and time, 6*, 11–58.

Staples, D. S. and Webster, J. (2008). Exploring the effects of trust, task interdependence, and virtualness on knowledge sharing and teams. *Information Systems Journal, 18*, 617–640.

Sweeney, P. J. (2007). Trust: The key to combat leadership. In D. Crandall (ed.), *Leadership lessons from West Point* (pp. 252–277). San Francisco: Jossey-Bass.

Usoro, A., Sharratt, M. W., Tsui, E., and Shekhar, S. (2007). Trust as an antecedent to knowledge sharing in virtual communities of practice. *Knowledge Management Research and Practice, 7*, 199–212.

Vogelaar, A. L. W. and Kramer E.-H. (2000). Mission command in ambiguous situations. In C. McCann and R. Pigeau (eds), *The human in command: Exploring the modern military experience* (pp. 217–231). New York: Kluwer Academic/Plenum Publishers.

Whitener, E. M., Brodt, S. E., Korsgaard, M. A., and Werner, J. M. (1998). Managers as initiators of trust: An exchange relationship framework for understanding managerial trustworthy behavior. *Academy of Management Review*, *23*(3), 513–530.

Xiao, Y., Seagull, F. J., Mackenzie, C., Ziegert, J., and Klein, K. (2003). Team communication patterns as measures of team processes: Exploring the effects of task urgency and shared team experience. *Proceedings of the 47th Annual Meeting of the Human Factors and Ergonomics Society* (pp.1502–1506). Santa Monica, CA: Human Factors and Ergonomics Society.

Zand, D. E. (1972). Trust and managerial problem solving. *Administrative Science Quarterly*, *17*(2), 229–239.

# Chapter 7
# Trust in International Military Missions: Violations of Trust and Strategies for Repair

Ritu Gill, Megan M. Thompson and Angela R. Febbraro
*Collaborative Performance and Learning Section, Defence Research & Development Canada, Toronto, Canada*

The need to better understand the nature and quality of the relationships between militaries and indigenous local populations in international missions has become apparent over recent years, in particular since militaries have entered more complex theatres of operations such as Kosovo, Bosnia, Africa and, more recently, Iraq and Afghanistan. Although it has almost always been the case that 'whenever a military force deploys somewhere, whether it is for a peace support operation, for humanitarian assistance, or for warfighting, there is always a civilian dimension to consider' (Longhurst, 2007, p. 57). Contemporary missions increasingly involve a variety of tasks ranging from warfighting through reconstruction and stabilization, providing additional opportunities for various kinds of interaction between a greater number of military personnel and members of the local population. The complexity and the often rapidly changing realities inherent in these missions also create increasing potential for ambiguity, uncertainty, errors and misunderstandings. One potentially vital aspect to consider in these missions is the role of trust (Gizewski and Rostek, 2007; Leslie, Gizewski, and Rostek, 2008; van der Kloet, 2006).

## Trust

Defined as the willingness to 'make oneself vulnerable to another party based on the expectation that the other will perform a particular action important to the trustor, irrespective of the ability to monitor or control that other party' (Mayer, Davis, and Schoorman, 1995, p. 712), trust has been termed the 'single most important element of a good working relationship' (Fisher and Brown, 1988, p. 107), and 'a foundation of social order' (Lewicki, McAllister and Bies, 1998, p. 438). Trust is comprised of four distinct, although positively related, dimensions: competence (belief in another's level of skill, ability and/or knowledge to accomplish a valued or needed task or goal); integrity (beliefs about another's honesty and adherence to high principles and codes of honor); benevolence (beliefs that the actions of other(s) are motivated by genuine care and concern); and predictability (beliefs that the reactions and behaviors of other(s) can be reliably anticipated – note however that predictability is not always included in conceptual models of trust) (Mayer et al., 1995; Stouffer, Adams, Sartori, and Thompson, 2008). Trust is assumed to be especially critical in situations that are characterized by interdependence, uncertainty, ambiguity, and risk (Holmes, 1991; Mayer et al., 1995; Mishra, 1996).

Theory and research suggest that trust promotes the ability to establish new and to maintain existing associations (Lewicki and Bunker, 1996; Mayer et al., 1995). Trust promotes positive performance outcomes (Davis, Schoorman, Mayer, and Tan, 2000; Dirks, 1999; Dirks and Ferrin, 2001), as it is associated with increases in the efficiency of information dissemination and processing (Krishnan, Martin, and Noorderhaven, 2006), reductions in concerns regarding sharing sensitive information, and increases in risk taking based on a stronger belief that a trusted person will 'watch my back' (Mayer and Davis, 1999). High trust generally reduces conflict in relationships (Dirks and Ferrin, 2001; Lewicki, 1995; Lewicki and Wiethoff, 2000; Massey and Dawes, 2007; Peterson and Behfar, 2003) and trusting partners tend to extend goodwill when unforeseen contingencies arise (Krishan et al., 2006), as well as to view equivocal actions in a constructive manner conducive to the stability of the relationship (Uzzi, 1997; Zaheer, McEvily, and Perrone, 1998). Moreover, trusting people spend less time monitoring the relationship and protecting against opportunistic behaviors (Bromiley and Cummings, 1995, Dore 1983). This means fewer 'demands for controls and protections against the possibility of betrayal (Tyler and Kramer, 1996, p. 4). Importantly, under crisis conditions, high trust promotes continued honest and open communication and the sharing of scarce resources (Mishra, 1996). Finally, because trust also is associated with greater feelings of shared identity and increased efforts to promote the welfare of the group, trust is associated with cooperative behaviors (Brewer and Miller, 1996) and the discovery of mutually beneficial, integrative solutions (Walton and McKersie, 1965). Conversely perceived violations of trust serve to undermine all the benefits cited. Indeed, violations may cause trust to plunge to a level that may be below the initial trust level, representing a particular challenge and requiring different strategies from those involved in initial trust development (Kim, Cooper, Ferrin, and Dirks, 2004).

Interestingly, recent organizational trust research (Kim, et al., 2004) found that trust was repaired more successfully when violators apologized (i.e., sincerely acknowledged both responsibility and regret for a violation) for errors associated with a lack of competency but denied responsibility for integrity violations (i.e., explicitly declared an allegation to be false, while assuming no responsibility or regret). The researchers explain these somewhat surprising findings by suggesting that individuals tend to weigh positive information regarding competence more heavily than negative information about competency, but conversely, tend to weigh negative information about integrity more heavily than positive information about integrity. Although an apology confirms that a violation was committed, the competence violation may be considered an anomaly that would not be repeated in the future, thus salvaging and repairing trust. In contrast, individuals may believe that those with high integrity will refrain from dishonest behavior, and those with low integrity may exhibit dishonest behavior. The authors suggest that apologizing for the integrity violation would be less effective than a denial, as the apology would confirm guilt and thus a lack of integrity and would have a negative effect on subsequent trust. Although this research is in its infancy and conducted within the context of organizational psychology, such studies provide a starting point for the exploration of trust violation and repair mechanisms in other contexts.

### *Trust in International Missions*

The development and maintenance of trust between military organizations and a local population is increasingly recognized as integral to the success of international military operations (e.g., Gizewski and Rostek, 2007; Leslie et al., 2008; van der Kloet, 2006). For instance, Gizewski and Rostek (2007) contend that future military operations are likely to be as much about gaining trust and legitimacy among the surrounding populations as engaging in armed combat and destroying adversaries. Certainly recent media coverage confirms the importance of gaining the trust and cooperation of the local population in order for militaries and international communities to attain their mission objectives. For instance, an interview with Canadian Defence Minister Peter MacKay and Canadian Forces Brigadier General Jonathan Vance indicated that the cooperation of Afghan villages is essential to the ability of Canadians to deliver development programs, including the immunization of children and the building of schools (Tibbetts, 2009). The Defence Minister further stated that there is a 'connection between our ability to deliver, and the security and the cooperation we require from local Afghans' (Tibbetts, 2009). US Army General Stanley A. McChrystal, former Commander of US Forces Afghanistan (USFOR-A) and of International Security Assistance Force (ISAF) in Afghanistan asserted that 'if the people are against us, we cannot be successful … if the people view us as occupiers and the enemy, we can't be successful and our casualties will go up dramatically' (CBC News, 2009). General McChrystal noted that it is important for him to convey a message of trust and to demonstrate that he was not more valuable than the Afghan citizens (CBC News, 2009) by not wearing body armour or carrying a sidearm when visiting Afghan governors and local Afghans.

Although support and trust are often initial responses of a local populace to a military presence, over time, a decrease in trust and support of the military by locals frequently occurs, as has been the case in Afghanistan (Security and Development Policy Group, 2007) and Bosnia (van der Kloet, 2006), to name but two examples. This shift in trust and support for military organizations can stem from acts of commission by the military (i.e., inflicting collateral damage), acts of omission on the part of the military (i.e., failing to deliver on promises), or because of misunderstanding – either by the military in terms of the needs of the local populace, or if the local population misunderstands the intent of actions by the military. Whatever the basis, these instances are often interpreted by the local population as a violation of their trust.

Within the context of international missions, real or perceived trust violations often strain the relationship between military organizations and the local population, undermining the military's ability to provide support and assistance to the local population and even encouraging the local population to give their support to local terrorist organizations instead (Security and Development Policy Group Council, 2007; van der Kloet, 2006), thus potentially jeopardizing mission success. For all of these reasons, understanding the dynamics of trust, trust violation, and trust repair in these contexts is vital. The current research is among the first to explore the issue empirically, in particular by applying recent results from the organizational trust violation and repair literature to the complex and increasingly important area of international counterinsurgency and stability missions. More specifically, based on the work of Kim and colleagues (2004), we sought to determine: (a) whether a particular trust repair strategy (apology vs. denial) was more effective after a trust violation and under which conditions this might be the case (competence vs. integrity trust violation); and (b) whether a particular trust repair strategy could ameliorate the effects of a trust violation relative to conditions under which no trust violation had occurred. Finally, given the lack of empirical data in this area, we also investigated the nature of repair strategies in the context of international missions in more detail by exploring participant's recommendations concerning optimal ways of regaining trust between a local population and a foreign military.

## Method

### *Participants*

Fifty civilians ($M$ age = 28.84, $SD$ = 11.10, 21 males and 29 females) participated in this study. The majority of participants were White ($n$ = 29, 58 percent), while the remainder of the sample was of Chinese ($n$ = 8, 16 percent), South Asian ($n$ = 6, 12 percent), Black ($n$ = 2, 4 percent), Filipino ($n$ = 2, 4 percent), Southeast Asian ($n$ = 2, 4 percent), and Aboriginal ($n$ = 1, 2 percent) descent. A large majority of participants were born in Canada ($n$ = 31, 62 percent), while the remainder were born in China, the US, Guyana, the Philippines, Hong Kong, Serbia, Singapore, Korea, India, Ukraine, Russia, or Romania (1 or 2 participants were born in each of these countries). The mean number of years that participants had resided in Canada was 23.26 ($SD$ = 11.41).

### *Procedure*

Participants were asked to imagine themselves as a local citizen of a fictional country, Safia, which was experiencing conflict due to the rising power of a local terrorist organization, in the context of a study assessing perceptions of the relationship between military organizations and the local population being assisted. An initial scenario section provided a historical background to the fictional country of Safia, including a summary of the issues related to the current conflict within the country, as well as background into the government of Massey (another fictional country) which offered its military to assist Safia in dealing with the terrorist threat and rebuilding the country, specifically by providing improved health care and road/travel conditions, an educational system, and political infrastructure (see Table 7.1). Baseline measures assessing expectations concerning the Massey military, perceptions of their competence, benevolence and integrity, and trusting intentions regarding the Massey military (see below) were then completed. Following from Kim et al's (2004) methodology, participants then read a second portion of the scenario that involved a competency-based violation, an integrity-based violation, or no trust violation. Participants were randomly assigned to trust violation (competency vs. integrity), trust repair (denial vs. apology) or control conditions (see Table 7.2), which were embedded within different versions of the scenario. The competence trust violation was constructed to indicate that the Massey military were unable to fulfill their mandate to provide basic resources and security to the local population due to their lack of knowledge/technical skills (see Butler and Cantrell, 1984), while the integrity trust violation was constructed to indicate that the Massey military failed to live up to their promises to provide security and resources because they appeared to be dishonest and self-serving (Mayer et al., 1995). In the control condition no trust violation was committed. Participants next completed a second set of dependent measures identical to the first, with the addition of items assessing their willingness to risk their future to the Massey military.

The next portion of the scenario contained the trust repair manipulation based on the Massey military's response to criticisms of their activities. In the apology repair condition, the Massey military admitted that mistakes had been made, took responsibility for the errors to date, admitted their culpability to date and promised to correct the errors in the future. In the denial repair condition the Massey military largely denied any responsibility for not delivering on their mandate within Safia – indeed they suggested that the criticism was largely political maneuvering on the part of the Safian government. The control condition included no trust repair manipulation as no trust violation had been committed. A third set of dependent measures identical to the second questionnaire administration was then completed in order to assess trust levels after the repair manipulation. Finally, participants provided short-answer recommendations concerning what they considered to be optimal trust repair strategies for this context, as well as manipulation checks.

**Table 7.1 Design overview**

| *Scenario introduction* | **Baseline measures** | *Trust violation manipulation* | **Post-violation measures** | *Trust repair manipulation* | **Post-repair measures** | **Post-scenario measures** |
|---|---|---|---|---|---|---|
| *Country history, background to current conflict and request by Safian Government for Massey military assistance* | a. Expectations of the Massey military<br>b. Trust of Massey military: (perceived competence, integrity, benevolence)<br>c. Trusting intentions toward military | *Integrity vs. competence vs. control* | a. Expectations of the Massey military<br>b. Trust of Massey military: (perceived competence, integrity, benevolence)<br>c. Trusting intentions toward Massey military<br>d. Willingness to risk aspects of future on Massey military | *Apology vs. denial vs. control* | a. Expectations of the Massey military<br>b. Trust of Massey military: (perceived competence, integrity, benevolence)<br>c. Trusting intentions toward Massey military<br>d. Willingness to risk aspects of future on Massey military | a. Trust repair recommendations (short answer)<br>b. Manipulation checks |

*Note*: All quantitative dependent measures were based on a seven-point scale (7 = strongly agree to 1 = strongly disagree).

**Table 7.2 Trust study conditions**

| | | **Trust violation** | | **Control** |
|---|---|---|---|---|
| | | Competence | Integrity | (i.e. No trust violation, No repair strategy) n = 10 |
| Trust repair strategy | Apology | n = 10 | n = 10 | |
| | Denial | n = 10 | n = 10 | |

The scenario and questionnaires were administered via computer, using MediaLab, excluding the Voluntary Consent Form, which was completed on paper. Only the researcher and participant had access to the room to ensure the privacy of participant responses. The session lasted approximately 1 hour. The methodology and materials were reviewed and approved by the Defence Research & Development Canada (DRDC) Toronto Human Research Ethics Committee.

*Measures*

*Expectations of the military* Participants answered seven items that assessed their expectations of the military organization assisting their country (e.g., 'The Massey military and local Safians are likely to work well together,' 'The Massey military is likely to have good training'), that were adapted from a measure developed by Brown, Adams, Famewo, and Karthaus (2008). Reliability analyses for the measure were excellent: baseline Cronbach's $\alpha = .81$, post-violation $\alpha = .93$, and post-repair $\alpha = .96$.

*Trust dimensions* A total of 17 items adapted from Mayer and Davis (1999) reflected the three most prominent trust dimensions. Reliabilities were all excellent (competence, six items: e.g., 'The Massey military are very capable of performing their job,' baseline $\alpha = .86$; post-violation $\alpha = .93$; post-repair $\alpha = .93$; benevolence, five items: e.g., 'The Massey military are very concerned about the welfare of Safians,' baseline $\alpha = .82$; post-violation $\alpha =. 93$; post-repair $\alpha = .96$; integrity, 6 items: e.g., 'The Massey military has a strong sense of justice,' baseline $\alpha = .84$; post-violation $\alpha = .92$; post-repair $\alpha = .93$).

*Trust intentions* Four items assessed participants' propensity to trust the military (e.g., 'I would let the Massey military continue to have influence over issues that are important to me for the duration of their mission here,' adapted from Mayer and Davis, 1999). This measure, administered three times, also had excellent reliabilities: baseline $\alpha = .81$, post- violation $\alpha = .91$, and post-repair $\alpha = .90$.

*Willingness to risk* Seven items measured the degree to which participants agreed with statements such as: 'If asked to be in a high-risk situation with the Massey military, I would be willing to rely on them to "watch my back"/take care of me' (adapted from Brown et al., 2008). This measure was administered twice and yielded excellent reliabilities: post-violation $\alpha = .96$; post-repair $\alpha = .96$.

*Open-ended trust repair strategy item* Participants were asked an open-ended question: 'As a local member of Safia, in order to gain your trust, what would you have liked the Massey military to have said or done?'

### *Manipulation Checks*

Participants answered three multiple-choice questions administered at the end of the study that assessed whether they recognized the nature of the violation (competence vs. integrity – two items) and the response to the violation (apology vs. denial – one item) (Kim et al., 2004).

## Results

### *Preliminary Analyses*

Cross-tabulation analyses presented in Table 7.3 indicate that over half of participants correctly identified the trust violation condition to which they were assigned. Average frequencies across the two questions ('In the scenario, the Massey military was accused of being unable to defeat the insurgency and improve conditions of Safia due to: _____' and 'What does this accusation bring into question?') ranged from 47 percent in the integrity trust violation to 65 percent in the competency violation and control conditions. Results for the trust repair manipulation check question ('What was the Massey military's response to the accusation?') strongly confirmed this aspect of our manipulations, with 90 percent of participants correctly identifying the denial, apology or no trust repair aspects within the scenario they had read.

Supporting the strength of our trust violation manipulations overall, there was a significant decrease in trust from baseline to post-violation in both the integrity and competence violation conditions (perceived competence, perceived integrity, and perceived benevolence, as well as trust intentions[1]; see Table 7.4). Thus, for the integrity violation: competence, $t(19) = 9.25$, $p < .001$, benevolence $t(19) = 5.99$, $p < .001$, integrity $t(19) = 5.39$, $p < .001$, and trust intentions $t(19) = 4.24$, $p < .001$; and for the competence violation: integrity $t(19) = 5.80$, $p < .001$, competence $t(19) = 8.77$, $p < .001$, benevolence $t(19) = 7.53$, $p < .001$, and trust intentions $t(19) = 6.36$, $p < .001$. Table 7.4 also shows that within the control condition, perceived competence ($t(9) = -1.85$, $p > .05$) and benevolence ($t(9) = -3.08$, $p > .05$) did not differ significantly between baseline and post-violation, although the pattern of means indicates that trust increased slightly baseline to post-violation. For integrity ($t(9) = -4.21$, $p < .05$) and trust intentions ($t(9) = -4.65$, $p < .05$), a significant difference was observed with an increase in trust from baseline to post-violation (see Table 7.4).

1 To minimize Type 1 errors, a Bonferroni adjustment was applied to all multiple comparison analyses to maintain a family-wise error rate of $\alpha = .05$.

**Table 7.3 Cross-tabulation results of trust violation and trust repair manipulation checks**

| | Percentage correctly identified |
|---|---|
| Manipulation checks | |
| Trust violation check #1* | |
| Competence trust violation condition | 60% |
| Integrity trust violation condition | 40% |
| Control condition | 70% |
| Trust violation check #2† | |
| Competence trust violation condition | 70% |
| Integrity trust violation condition | 55% |
| Control condition | 60% |
| Trust repair check‡ | |
| Apology repair condition | 90% |
| Denial repair condition | 90% |
| Control condition | 90% |

* Participants were asked: 'In the scenario, the Massey military was accused of being unable to defeat the insurgency and improve conditions of Safia due to?'

† Participants were asked: 'What does this accusation bring into question?'

‡ Participants were asked: 'What was the Massey military's response to the accusation?'

**Table 7.4 Means and standard deviations of trust dimensions (perceived competence, integrity, benevolence) and trust intentions by trust violations**

| | Trust violations | | | | | |
|---|---|---|---|---|---|---|
| | Competence | | Integrity | | Control | |
| | Mean | (SD) | Mean | (SD) | Mean | (SD) |
| Perceived competence | | | | | | |
| Baseline | 4.86[a] | (0.96) | 5.51[a] | (0.61) | 4.58 | (1.16) |
| Post-violation | 2.29[b] | (0.96) | 3.34[b] | (1.04) | 5.25 | (0.87) |
| Perceived benevolence | | | | | | |
| Baseline | 4.41[a] | (1.01) | 4.11[a] | (0.94) | 4.02 | (1.06) |
| Post-violation | 2.47[b] | (0.87) | 2.41[b] | (1.15) | 5.16 | (1.19) |

**Table 7.4** ***Concluded***

| | Trust violations | | | | | |
|---|---|---|---|---|---|---|
| | Competence | | Integrity | | Control | |
| | Mean | (SD) | Mean | (SD) | Mean | (SD) |
| Perceived integrity | | | | | | |
| Baseline | $4.27^a$ | (0.94) | $4.18^a$ | (0.93) | $4.11^a$ | (0.91) |
| Post-violation | $2.63^b$ | (0.93) | $2.75^b$ | (1.01) | $5.23^b$ | (0.74) |
| Trust intentions | | | | | | |
| Baseline | $3.58^a$ | (0.91) | $4.03^a$ | (1.06) | $3.30^a$ | (1.14) |
| Post-violation | $1.95^b$ | (0.91) | $2.68^b$ | (1.20) | $4.90^b$ | (1.16) |

*Note*: Columns with different superscripts differed at p<.05

Correlational analyses presented in Table 7.5 also supports previous trust findings in that baseline competency was moderately correlated with integrity and benevolence ($r$'s = .41 and .45, respectively, both $p$'s < .001), although integrity and benevolence were highly positively correlated with each other ($r = .75$, $p < .001$). Interestingly, however, at both the post-violation and post-repair assessments, the three trust dimensions were very highly correlated (average $r$'s = .84 and .77, respectively, $p$'s < .001).

**Table 7.5** **Correlations between trust dimensions baseline, post-trust violation, and post-trust repair**

| | 1 | 2 | 3 | 4 | 5 | 6 | 7 | 8 | 9 |
|---|---|---|---|---|---|---|---|---|---|
| Trust dimensions: Baseline | | | | | | | | | |
| 1. Perceived competence | - | $.45^{**}$ | $.41^{**}$ | .01 | -.08 | -.08 | .10 | -.07 | -.14 |
| 2. Perceived integrity | - | - | $.75^{**}$ | .01 | .11 | .18 | .01 | .15 | .06 |
| 3. Perceived benevolence | - | - | - | -.03 | -.19 | .14 | .01 | .05 | .06 |
| Trust dimensions: Post-Violation | | | | | | | | | |
| 4. Perceived competence | - | - | - | - | $.84^{**}$ | $.78^{**}$ | $.86^{**}$ | $.77^{**}$ | $.66^{**}$ |
| 5. Perceived integrity | - | - | - | - | - | $.89^{**}$ | $66^{**}$ | $.83^{**}$ | $.78^{**}$ |
| 6. Perceived benevolence | - | - | - | - | - | - | $.68^{**}$ | $.82^{**}$ | $.81^{**}$ |
| Trust dimensions: Post-repair | | | | | | | | | |
| 7. Perceived competence | - | - | - | - | - | - | - | $.75^{**}$ | $.66^{**}$ |
| 8. Perceived integrity | - | - | - | - | - | - | - | - | $.90^{**}$ |
| 9. Perceived benevolence | - | - | - | - | - | - | - | - | - |

*Note*: $^{*}p < .05$, $^{**}p < .001$

*Effects of Trust Violations and Trust Repair Strategies on Intentions to Trust, Future Willingness to Risk, and Positive Expectations of Future Assistance*

Next, we addressed the core questions of this research by conducting three univariate 2 (trust violation: competency vs. integrity) × 2 (trust repair: apology vs. denial) ANOVAS on the dependent variables of trust intentions, willingness to risk, and expectations concerning the Massey military, respectively.[2] As anticipated given the results of the manipulation checks, participants who did not experience a trust violation (control condition) reported a significantly greater propensity to trust the military than those who experienced the competence and integrity trust violations, $F(1,45) = 4.59$, $p < .05$, $\eta^2 = .093$ (see Table 7.6). In addition, participants in the control condition also had a tendency to a greater propensity to trust the military compared to those who experienced an apology or denial trust repair strategy, $F(1,45) = 5.54$, $p < .05$, $\eta^2 = .110$.

**Table 7.6 Means and standard deviations of trust intentions by trust violations and trust repair**

| | Trust violation | | | |
|---|---|---|---|---|
| | Competence | | Integrity | |
| | Mean | (SD) | Mean | (SD) |
| Trust repair | | | | |
| Apology | $2.62^{a,c}$ | (1.36) | $3.40^{a,c}$ | (1.23) |
| Denial | $1.80^{a,c}$ | (0.68) | $2.55^{a,c}$ | (1.27) |
| Control | $4.68^{b,d}$ | (0.93) | $4.68^{b,d}$ | (0.93) |

*Note*: Columns with different superscripts differed at p<.05

Moreover, participants in the control condition were more willing to risk (i.e., put their own life on the line), and were more likely to endorse a belief that the military will 'watch their back,' than were those who experienced an apology or denial trust repair strategy, $F(1,45) = 14.54$, $p < .001$, $\eta^2 = .244$. However, follow-up comparisons also revealed that those who experienced an apology were also significantly more willing to risk compared to those who experienced a denial trust repair strategy (see Table 7.7).

2 ITo minimize Type 1 errors, a Bonferroni adjustment was applied to all multiple comparisons analyses to maintain a family-wise error rate of $\alpha = .05$.

**Table 7.7 Means and standard deviations of willingness to risk by trust violations and trust repair**

| | Trust violation | | | |
|---|---|---|---|---|
| | Competence | | Integrity | |
| | Mean | (SD) | Mean | (SD) |
| Trust repair | | | | |
| Apology | $3.27^{a,c}$ | (0.91) | $3.17^{a,c}$ | (1.16) |
| Denial | $1.90^{a,d}$ | (0.83) | $2.18^{a,d}$ | (1.02) |
| Control | $4.87^{b}$ | (0.91) | $4.87^{b}$ | (0.91) |

*Note*: Columns with different superscripts differed at $p<.05$

Follow-up comparisons also indicated that participants who did not experience a trust violation (control condition) reported significantly more positive/higher expectations of the military assisting their country than did those who experienced an apology or denial trust repair strategy (see Table 7.8). Still, those individuals who experienced an apology had significantly more positive expectations of the military assisting their country compared to those who experienced a denial, $F(1,45) = 6.23$, $p < .05$, $\eta^2 = .122$ (see Table 7.8).

**Table 7.8 Means and standard deviations of expectations by trust violations and trust repair**

| | Trust violation | | | |
|---|---|---|---|---|
| | Competence | | Integrity | |
| | Mean | (SD) | Mean | (SD) |
| Trust repair | | | | |
| Apology | $3.32^{a,c}$ | (0.93) | $3.31^{a,c}$ | (1.10) |
| Denial | $2.20^{a,d}$ | (0.91) | $3.00^{a,d}$ | (0.94) |
| Control | $5.17^{b}$ | (0.58) | $5.17^{b}$ | (0.58) |

*Effects of Trust Violations and Repair Strategies on Trust Dimensions (Perceived Competence, Integrity, and Benevolence)*

A MANOVA was conducted on trust dimensions (perceived competence, perceived integrity, and perceived benevolence) in order to assess the conditions under which trust is most likely to be repaired. A significant multivariate main effect was observed for the trust violation condition, Pillais = .286, $F(3,43) = 5.73$,

$p < .05$, $\eta^2 = .286$, and follow-up univariate analyses indicated that participants who did not experience a trust violation (control condition) perceived the military to have greater competence compared to those who experienced a competence or integrity trust violation, $F(1,45) = 12.83$, $p < .05$, $\eta^2 = .222$. In addition, follow-up comparisons showed that those who experienced the integrity trust violation perceived greater competence in the military compared to those who experienced the competence trust violation (see Table 7.9).

**Table 7.9 Means and standard deviations of trust dimensions (perceived competence, perceived integrity, perceived benevolence) by trust violations and trust repair**

| | Trust violation | | | | | |
|---|---|---|---|---|---|---|
| | Competence | | Integrity | | Control* | |
| | Mean | (SD) | Mean | (SD) | Mean | (SD) |
| Trust repair | | | | | | |
| Apology | | | | | | |
| Perceived competence | $3.08^{a,c}$ | (1.16) | $3.88^{a,d}$ | (1.17) | $5.45^{b}$ | (0.52) |
| Perceived integrity | $3.31^{a,c,e,g}$ | (0.89) | $3.41^{a,d,e,g}$ | (0.97) | $5.25^{b}$ | (0.56) |
| Perceived benevolence | $3.48^{a,d,e,g}$ | (1.10) | $3.12^{a,d,e,g}$ | (0.89) | $5.14^{b}$ | (0.73) |
| Denial | | | | | | |
| Perceived competence | $2.31^{a,c}$ | (0.92) | $3.91^{a,d}$ | (1.32) | $5.45^{b}$ | (0.52) |
| Perceived integrity | $2.07^{a,c,e,g}$ | (0.93) | $2.68^{a,d,e,g}$ | (0.98) | $5.25^{b}$ | (0.56) |
| Perceived benevolence | $1.92^{a,c,e,g}$ | (0.84) | $2.20^{a,d,e,g}$ | (1.20) | $5.14^{b}$ | (0.73) |
| Control* | | | | | | |
| Perceived competence | $5.45^{f,h}$ | (0.52) | $5.45^{f,h}$ | (0.52) | | |
| Perceived integrity | $5.25^{f,h}$ | (0.56) | $5.25^{f,h}$ | (0.56) | | |
| Perceived benevolence | $5.14^{f,h}$ | (0.73) | $5.14^{f,h}$ | (0.58) | | |

*Note*: Superscripts a to d denote significant differences in a given column at p<.05. Superscripts e to h denote significant differences in a given row at p<.05

* This group acted as a control in which participants do not experience the trust violation or trust repair manipulations.

There was also a significant multivariate main effect for the trust repair condition, Pillais = .291, $F(3,43) = 5.89$, $p < .05$, $\eta^2 = .291$, and univariate analyses indicated that participants in the control condition perceived the military to possess greater benevolence and greater integrity compared to those who experienced the apology or denial trust repair strategy. However, it was also the case that those who experienced the apology trust repair strategy perceived the military to possess greater benevolence and greater integrity compared to those who experienced the denial trust repair strategy, $F(1,45) = 16.36, p < .001, \eta^2 = .267$, and $F(1,45) = 12.53, p < .05, \eta^2 = .218$, respectively (see Table 7.9).

*Participant Recommendations for Trust Repair*

We also explored participants' recommendations for trust restoration in this context and whether these spontaneously generated recommendations would reliably reflect the four major theoretical dimensions of trust (competence, integrity, benevolence, and predictability). Sentences of the short-answer responses were coded using a nine-category coding scheme (i.e., positive and negative valence for each of the four trust dimensions, and a miscellaneous category) by two independent coders using NVivo8 (QSR International's NVivo8, 2008), a qualitative research data-analytic software package. Sentences that referred to more than one theme (e.g., referred to both benevolence and competence) could be coded as reflecting more than one category. The inter-rater reliability (overall mean kappa) was 0.76, which is considered to be excellent agreement (Capozzoli, McSweeney, and Sinha, 1999), and the percent agreement between the raters was 95.4 percent.

Interestingly, and somewhat unexpectedly given that the trust violations were constructed to reflect integrity and competence, 50 percent of participant-generated recommendations for trust restoration in this context were categorized as reflecting benevolence issues. These included recommendations that the military demonstrate 'more concern for improvements in health and life conditions' of the Safians, 'look for ways to help out the local economy and businesses,' 'protect and [improve] the lives of our [Safian] people,' 'keep the local population the priority in all decision making,' and invest more effort in ensuring that living conditions for local citizens are improved. In addition, 26 percent of participants' responses were categorized into the perceived competence trust dimension, with several participants commenting upon the importance of the military's technical ability and ability to effectively communicate progress or delays to the locals for improving trust.[3] For instance, participants indicated the need for the military to have more training to accomplish goals, to provide more stability for the locals (e.g., job opportunities, resources, security), and to provide to the locals a detailed explanation of problems and a plan to overcome those problems in rebuilding Safia. A further 20 percent of participants' responses were categorized as reflecting integrity-based issues. Several responses cited the importance of the Massey military taking responsibility for their faults, admitting their mistakes, fulfilling promises, and being honest in order to improve trust between the military and local population.

3 These percentages are on the low side compared to kappa and percentage agreement because they refer only to text that was mutually coded by coders, whereas kappa and percent agreement includes text mutually coded by coders and agreement by coders of text not to be coded in a category.

## Discussion

Future international missions are expected to have increasing potential for complexity, ambiguity, and rapidly changing circumstances. They are also anticipated to increasingly entail greater interaction between deployed militaries and members of local communities. Trust has been hypothesized as one of the key psychosocial enablers of effective interactions in these environments, supporting overall mission success in these fluid and challenging contexts (CBC News, 2009; Gizewski and Rostek, 2007; van der Kloet, 2006). Despite this emerging acknowledgement, to date little empirical research has explored this dynamic within international missions. Thus, our research represents one of the first efforts in this area. Specifically, within the context of international military engagements, we apply recent trust violation and repair findings from within the organizational trust literature (Kim et al., 2004) that suggested that denial is a more effective repair strategy for integrity-based trust violations and that apology is a more effective trust repair strategy for a competency-based trust violation.

To that end, 50 civilian participants read a fictional scenario with elements taken from recent international missions. Asked to take the perspective of a member of the local community, participants first read scenario elements that provided the background and history of the country (Safia), the basis of the current conflict, as well as the call for and agreement concerning foreign military assistance. After making initial trust assessments, participants continued to read the scenario in which the military committed a series of trust violations: 20 participants read scenarios indicating competency-based violations, 20 participants read scenarios indicating integrity-based violations, and 10 control participants read scenarios in which no violations occurred. A variety of trust measures were administered at this point. Experimental participants then read a scenario segment in which the military response was either an apology or denial of responsibility for problems within the international mission (control subjects did not experience a trust repair strategy as no violation had occurred) and trust measures were again administered. Once the scenario was completed, respondents also provided short-answer responses indicating their recommendations for trust remediation in this context.

Overall, the quantitative results are largely consistent with the previous trust literature. For instance, trust impairment followed trust violations. Indeed, decreases in trust after trust violation were observed across all trust measures, including the three trust dimensions (perceived competence, integrity, and benevolence) and trust intentions, regardless of the trust violation condition. For instance, in the competence trust violation condition, a decrease in trust was observed not only for the competence trust dimension, but also for the integrity and benevolence trust dimensions. Likewise, for the integrity trust violation condition, a decrease in trust was observed not only for the integrity trust dimension, but also for the competence and benevolence trust dimensions. Further, our results showed that some recovery of trust occurred, at least to a limited extent following the trust repair attempts.

Also consistent with theorizing in this area, our results showed that the best trust-related outcomes occur in the absence of trust violations. However, in contrast to previous research (e.g., Kim et al., 2004), where trust violations did occur, it appears that the most effective trust repair strategy is an apology rather than a denial, although again it is important to note that the means for trust measures following either repair strategy remained low. Notably, while Kim and colleagues found a significant interaction between the trust violation and trust repair manipulations (e.g., apology more effective for competence violation, denial more effective for integrity violation), we found a significant main effect for the trust repair manipulation (e.g., apology more effective than denial regardless of type of trust violation). This difference may be attributed to the fact that a significant difference between these two studies is context; our study examines trust violations and trust repair in a local population from a war-torn country, whereas Kim and colleagues examine these trust issues specifically within an organizational setting (i.e., a hiring decision). These two contexts, and populations, face very different challenges in terms of the magnitude and implications of the trust violations, and there were obviously varying repercussions for restoring trust. However, it may also be the case that some of these differences might be associated with our study being preliminary in nature and having low sample size and power. In continuing further studies with greater power we will be able to determine the nature of any discrepant results observed.

The general pattern of results here is also consistent with past literature concerning the hypothesized positive interrelationships among the dimensions that comprise trust (Mayer et al., 1995). This pattern of moderately positive correlations among trust dimensions was most apparent at baseline, although there a higher magnitude of correlations existed between integrity and benevolence – a point to which we will return later in the discussion. Also of interest, we saw the magnitude of the correlations among trust dimensions rise dramatically after the trust violation and post-repair manipulations. As the items reflecting the different trust dimensions did not change pre- to post-violation, perhaps this finding reflects something about the dynamics of trust when violated – that is, mistrust in one dimension 'leaks' to the others – as well as about the dynamics of trust repair, at least in the short term. Further studies are needed to confirm this pattern of results as well as to explore the psychological dynamics that may account for these findings.

In addition to quantitative findings, qualitative analyses revealed that the major trust dimensions are indeed reflected in participants' recommendations for ways to restore trust, with each dimension representing an important component in gaining trust. Interestingly, benevolence appears to be the most often cited factor in restoring trust, with half of participants suggesting that the military provide basic needs for the local population and demonstrate greater concern for the Safians through improvements in health and life conditions.

As this study is one of the first to use this methodology to approach the issue of trust within this context, the current study, while valuable, does have some limitations which could impact upon the generalizability of these results to real international missions. For instance, although we took care to develop trust violations and repair that were based on factual incidents from a variety of recent missions, the scenario-based design means that these elements are not comparable to real-life experiences of coming from a war-torn country. The latter involves intense and chronic stress and grave consequences, as well as a lack of resources to survive. In the current study, participants provided their responses without the emotional surround of these factors; although asked to imagine themselves as members of the local population and react accordingly, the actual negative consequences of experiencing the trust violations and trust repair strategies were minimal at best and certainly did not extend beyond the laboratory or session. In addition, the study would have been strengthened with the use of a larger number of respondents. Nonetheless, despite a relatively small sample size and low statistical power, the presence of significant results suggests robust findings.

Furthermore, the results of the manipulation check analyses indicated some confusion concerning experimental condition assignment, in particular within the integrity trust violation condition. Moreover, baseline correlations between integrity and benevolence were quite high, and remained so post-violation and post-repair. Upon reflection, embedded within the integrity violation condition were elements often tied not only to dishonesty of the Massey military but also to their taking advantage of the situation for their own gain and comfort, suggesting not only a lack of integrity but also a lack of benevolence. Although the post-violation and post-repair correlations might reflect the need to refine the integrity-related aspects of this scenario in this respect, it does not explain the higher correlation between integrity and benevolence at the baseline assessment. Thus, this finding may also relate to the theorized inter-relationship among the dimensions of trust. The latter rationale may have additional traction in light of our other results showing dramatic increases in the correlations among all of the trust dimensions after a trust violation, a relationship that remained even after the trust repair manipulations occurred. At this point the underlying reasons for the results are still speculation and it is left to future research to address this intriguing issue. Likewise, the reasons for the discrepancy between the present findings and Kim et al.'s (2004) findings regarding effective trust repair strategies would also be worth pursuing in future research.

Given this specific area, future research should also examine trust violation and repair in the context of a more culturally diverse sample. That is, although our sample had a varied cultural background, the vast majority of participants had lived in Canada all or most of their lives. Thus, it would also be interesting and indeed important to repeat the study with a sample that was more truly culturally diverse, and which perhaps included more recent arrivals to Canada, given the cultural diversity in the lands to which Canada and many Western countries deploy on expeditionary operations.

Given the limitations of this study, the results presented here are not intended to provide a basis for implementing any policies or procedures at this stage.

Certainly, the first order of business is to replicate the study with a larger sample to determine the stability of the findings. Indeed, we are currently conducting such studies within our laboratory. In addition, in actual field settings, trust violation and repair responses are likely to be influenced by factors such as the social context and culture, the quality of existing relationships, relationship history, and the reputations of individuals involved, all of which speak to the need for an integrated program of field and laboratory research that tests increasingly complex models. That being said, a distinct advantage of studying trust in the current context is the direct manipulation of trust variables in the context of the control provided within laboratory conditions. Thus, despite the preliminary nature of these findings, they do provide an informative starting point for the examination of the critical issue of trust violation and restoration within the context of an international military engagement.

## References

Brewer, M. B. and Miller, N. (1996) *Intergroup Relations*. Brooks/Cole Publishing: Pacific Grove CA.

Bromiley, P. and Cummings, L. L. (1995) Transaction costs in organizations with trust. In R. B. Bies, B. Sheppard, and R. Lewicki (eds), *Research on Negotiations in Organizations* (pp. 219–247). JAI Press: Greenwich, CT.

Brown A. L., Adams, B. D., Famewo J. J., and Karthaus, C. L. (2008) *Trust in Culturally Diverse Teams*. DRDC-Toronto-CF-2008–097.

Butler, J. K., Jr. and Cantrell, R. S. (1984) A behavioral decision theory approach to modeling dyadic trust in superiors and subordinates. *Psychological Reports*, 55, 19–28.

Capozzoli, M., McSweeney, L., and Sinha, D. (1999) Coefficient kappa: A review of interrater agreement. *Canadian Journal of Statistics*, 27(1), 3–23.

CBC News (2009, September 21) Harper holds firm to Afghanistan pullout date. Accessed from http://www.cbc.ca/world/story/2009/09/21/mcchrystal-afghanistan-troops-report901.html

Davis, J. H., Shoorman, F.D., Mayer, R.C., and Tan, H.H. (2000) The trusted general manager and business unit performance: Empirical evidence of a competitive advantage. *Strategic Management Journal*, 21, 563–576.

Dirks, K. T. (1999) The effects of interpersonal trust on work group performance. *Journal of Applied Psychology*, 84, 445–455.

Dirks, K. T. and Ferrin, D.L. (2001) The role of interpersonal trust in organizational settings. *Organization Science*, 12, 450–467.

Dirks, K. T. and Ferrin, D. L. (2002) Trust in leadership: Meta-analytic findings and implications for organizational research. *Journal of Applied Psychology*, 87, 611–628.

Dore, R. (1983) Goodwill and the spirit of market capitalism. *British Journal of Sociology*, 34, 459–82.

Fisher, R. and Brown, S. (1988) *Getting Together: Building Relationships as We Negotiate*. Penguin: London.
Gizewski P. and Rostek, M. (2007) Toward a JIMP-capable Land Force. *Canadian Army Journal*, 10(1), 55–72.
Holmes, J.G. (1991) Trust and the appraisal process in close relationships. In W. H. Jones and D. Perlman (eds), *Advances in Personal Relationships* (pp. 57–104). Jessica Kingsley: London.
Kim, P. H., Cooper C. D, Ferrin D. L., and Dirks, K. T. (2004) Removing the shadow of suspicion: The effects of apology versus denial for repairing competence- versus integrity-based trust violations. *Journal of Applied Psychology*, 89(1), 104–118.
Krishnan, R., Martin, X., and Noorderhaven, N. G. (2006) When does trust matter to alliance performance? *Academy of Management Journal*, 49, 894–917.
Leslie, A., Gizewski, P., and Rostek, M. (2008) Developing a comprehensive approach to Canadian Forces operations. *Canadian Military Journal*, 9(1), 11–20.
Lewicki, R. J. (1995) Trust in relationships: A model of development and decline. In M. Deutsch (ed.), *Conflict, Cooperation and Justice: Essays Inspired by the Work of Morton Deutsch*. Jossey-Bass: San Francisco.
Lewicki, R. J and Bunker, B. B. (1996) Developing and maintaining trust in work relationships. In R. M. Kramer and T. Tyler (eds), *Trust in Organizations: Frontiers of Theory and Research*, 114–139. Sage Publications: Thousand Oaks, CA.
Lewicki, R. J., McAllister, D. J., and Bies, D. J. (1998) Trust and distrust: New relationships and realities. *The Academy of Management Review*, 23, 438–458.
Lewicki, R. J. and Wiethoff, C. (2000) Trust, trust development and trust repair. In M. Deutsch and P. T. Coleman (eds), *The Handbook of Conflict Resolution: Theory and Practice*. Jossey-Bass: San Francisco.
Longhurst, G. M. (2007) The Evolution of Canadian Civil-Military Cooperation (CIMIC) *Canadian Military Journal*, Winter, 55–64.
Mair, S. (2008) The need to focus on failing states (2009). *Failed States*, 29, http://hir.harvard.edu/index.php?page = articleandid = 1708 (accessed January 2009).
Massey, G. R. and Dawes, P. L. (2007) Personal characteristics, trust, conflict, and effectiveness in marketing/sales working relationships. *European Journal of Marketing*, 41, 1117–1145.
Mayer R. C., Davis, J. H., and Schoorman, D. F. (1995) An integrative model of organizational trust. *Academy of Management Review*, 20(3), 709–734.
Mayer R. C. and Davis, J. H. (1999) The effect of the performance appraisal system on trust for management: Field quasi-experiment. *Journal of Applied Psychology*, 84(1), 123–126.
Mishra, A. K. (1996) Organizational responses to crisis: The centrality of trust. In R. M. Kramer and T. R. Tyler (eds), *Trust in Organisations* (pp. 261–287). Sage: London.

Nvivo qualitative data analysis software; QSR International Pty Ltd. Version 8, 2008.

Peterson, R. S. and Behfar, K. J. (2003) The dynamic relationship between performance feedback, trust, and conflict in groups. *Organizational Behaviour and Human Decision Processes*, 92, 102–112.

Stouffer, J., Adams, B., Sartori, J., and Thompson, M. (2008) Trust. In B. Horn and R. W. Walker (eds), *The Military Leadership Handbook* (pp. 525–542). Canadian Defence Academy Press/Dundurn Press: Kingston/Toronto.

Security and Development Policy Group: The Senlis Council (2007) *Peace in Afghanistan – Made in Canada.*

Tibbetts J. (2009, September 29) General's anger justified, MacKay says; Commander of Canadian troops lashes out at villagers after explosion. *Montreal Gazette*. Accessed from http://www.montrealgazette.com/news/2045128/story.html.

Tyler, T. R. and Kramer, R. M. (1996) Wither trust? In R. M. Kramer and T. R. Tyler (eds), *Trust in Organizations: Frontiers of Theory and Research* (pp. 1–15). Sage: Thousand Oaks, CA.

Uzzi, B. (1997) Social structure and competition in interfirm networks: The paradox of embeddedness. *Administrative Science Quarterly*, 42, 35–67.

Van der Kloet, I. (2006) Building trust in the mission area: A weapon against terrorism? *Small Wars and Insurgencies*, 17(4), 421–436.

Walton, R.E. and McKersie, R.B. (1965) *A Behavioural Theory of Labor Negotiations: An Analysis of a Social Interaction System*. McGraw-Hill: New York.

Zaheer, A., McEvily, B., and Perrone, V. (1998) Does trust matter? Exploring the effects of interorganizational and interpersonal trust on performance. Organization Science, 9, 141–159.

# Chapter 8
# Cultural Influences on Trust

Sandra C. Hughes, Cecily E. E. McCoy, Gabriella Severe and Joan H. Johnston

*Naval Air Warfare Center Training Systems Division, Orlando, FL, USA*

## Introduction

The United States Department of Defense (DOD) has recognized the critical need to improve Cross-Cultural Competence (3C) in military personnel (DOD Directive 1322.18, January 2009). 3C is the degree to which a person holds a set of characteristics (e.g., knowledge, skills, abilities, and other characteristics) that allow for the individual to successfully interact in any culturally complex environment to achieve the desired outcome, without necessarily having prior experience with the particular group, region, or language in that environment (Abbe, Gulick, and Herman, 2007; McCoy and Severe, 2010). 3C is a generalized cultural competence that is often contrasted with region-specific knowledge. Military services in the US are working together to create guidelines for 3C education, training, assessment, and practices (see McDonald, McGuire, Johnston, Selmeski, and Abbe, 2008). The 3C requirement stems from the changing nature of US military operations. Instead of traditional warfare, we are now more likely to engage in stability operations abroad, which focus on: reestablishing a safe and secure environment, providing essential governmental services, reconstructing emergency infrastructure, and providing humanitarian relief (Department of the Army, 2008). These missions require direct interactions between our military personnel and local civilians, as well as local security forces. When possible, it is best to obtain cooperation and support from local communities by establishing goodwill, rather than using force. Instead of being seen as an occupying force, we want to be seen as a resource providing help. Trust is often viewed as the most important component of a good working relationship, regardless of the context (Fisher and Brown, 1988). In stability operations, mission success often depends on military personnel's ability to develop interpersonal trust with individuals in host nations, because they need their cooperation and support. However, trust may be difficult to obtain in missions where host nationals may initially see the US military as an extremely distant culture, at best, or an adversarial, occupying force with goals that are mutually exclusive to their own, at worst. For this reason, the authors view the knowledge and skills associated with trust building in a multi-cultural context as a key component of 3C.

A recent Air Force report (Hardison, Sims, Ali, Villamizar, Mundell, and Howe, 2009) identified 69 behaviors required for effective cross-cultural interactions when deployed. Many of these behaviors are directly related to trust. They suggest that Air Force personnel need to know how to gain trust, respect, and credibility with locals. Further, they may need to change people's negative perceptions of Americans and the US military. In a similar vein, the Army's Field Manual on Stability Operations discusses the importance of positively influencing the opinions and attitudes of the host-nation population through information engagement, presence, and conduct (Department of Army, 2008). Moreover, it suggests that influence is best maintained by exhibiting respect for, and operating within, the cultural and societal norms of the local populace.

The goal of this chapter is to extend an established model of trust (Mayer, Davis, and Schoorman, 1995) to include cultural factors that guide hypotheses about trust development in an international context. A scenario is used to illustrate how specific cultural components can influence trust building, in a situation where a US serviceperson is trying to gain the trust of local leaders. Finally, practical implications for building trust relationships in multicultural interactions are discussed and recommendations for future research are provided.

## Trust

Mayer et al. define trust as 'the willingness of a party to be vulnerable to the outcomes of another party based on the expectation that the other will perform a particular action important to the trustor, irrespective of the ability to monitor or control that other party' (p. 712). Trust is only essential when certain mission/task conditions exist. For example, trust is necessary when both parties work *interdependently* and when there is *risk*, *uncertainty,* and *vulnerability* (North Atlantic Treaty Organization [NATO], in press). *Interdependence* is necessary because without being connected with another person, and without having one's outcomes in some way dependent on another person, there is no need to trust (Lewis and Weingert, 1985). *Risk* is a condition of trust because the trustor is not guaranteed a positive outcome. Rousseau, Sitkin, Burt, and Camerer (1998) define *risk* as the perceived probability of loss, as interpreted by a potential trustor. Mayer et al. distinguish between trust and cooperation, noting that cooperation can occur without trust, but cooperation that requires *risk* takes trust. *Uncertainty* occurs when the motives, goals, and future actions of others are not usually fully known. It is also an important precursor to trust because if one were omniscient, actions could be undertaken with complete certainty, leaving no need, or even possibility, for trust to develop (Lewis and Weingert, 1985). *Vulnerability* involves the possibility for the trusted party to take advantage of the trustor. If there is no vulnerability, then there is no need for trust.

Figure 8.1 presents an adaptation of the Mayer et al. input–process–output model of trust—The Intercultural Model of Trust (IMT). The original model (represented by the gray boxes in the figure) has been widely accepted in the literature and supported by numerous empirical studies (e.g., Branzei, Vertinsky, and Camp, 2007; Tzafrir, 2005). According to Mayer, et al., (1995) the major inputs to trust are *cognitions* that influence trust. They include the trustor's perceptions about the trustee's *ability*, *benevolence*, and *integrity*. A trustor's *perceptions about ability* include judgments about the trustee's skills, competencies, and characteristics related to the specific situation. *Perceptions about benevolence* are judgments about the trustee's kindness and goodwill towards the trustor. They also include loyalty, receptivity, and caring. *Perceptions about integrity* pertain to judgments about the trustee's consistency, fairness, reliability, and openness. In addition, judgments about trustor-trustee value similarity is a component of perceived integrity.. Mayer et al. suggest that positive perceptions about all three components must occur in order for trust to develop. Clearly, these perceptions are necessary for a trustor to make accurate predictions about a trustee. Another factor that should be considered is the trustor's propensity to trust; people simply differ in the extent to which they are willing to trust (Mayer et al.). Thus, according to Mayer et al., perceptions, along with dispositional factors, are inputs to trust development.

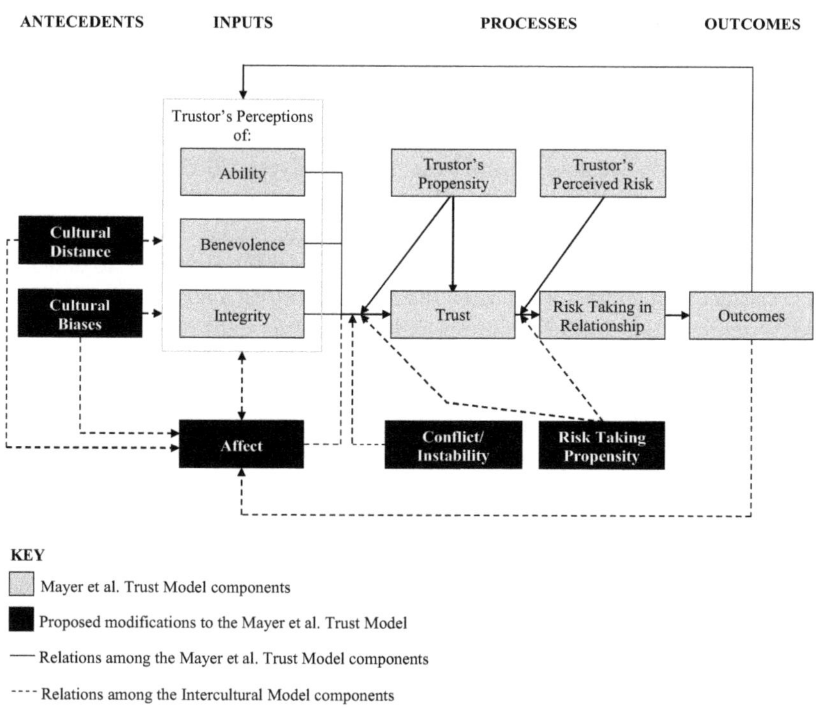


**Figure 8.1 The Intercultural Model of Trust (IMT)**

After initial trustworthiness is established, the trustor weighs the *perceived risks* and benefits before engaging in the actual risk-taking behavior. If the risks outweigh the benefits, the trustor may avoid risk-taking behaviors even if s/he perceives the trustee as trustworthy. This relation varies due to the level of the trustor's risk-taking propensity, with risk-seekers more likely to engage in the risk-taking behaviors than risk-averse individuals. However, if the trustor decides to take the risk, the decision can lead to positive and/or negative *outcomes.* Finally, the feedback loop recycles back into the inputs, where the *cognitions* update, and the trust-building process renews. Over time, repeated interactions help solidify judgments about the trustee's trustworthiness (Ting-Toomey and Oetzel, 2001) and influence future trust decisions.

The empirical basis for the Mayer et al. model is based on Western notions of trust, thus representing a potential gap in our understanding of the development of trust across cultures. The authors believe that the model is relevant across cultures because it is premised upon principles of human behavior and cognition. However, few empirical studies prescribe trust development in multicultural environments. This chapter takes the first step in developing a theoretical framework that proposes relations between culture and the trust process.

## Culture

Culture represents the enduring behaviors, ideas, attitudes, symbols, and traditions shared by a large group of people and transmitted from one generation to the next (Matsumoto, 2009; Matsumoto and Juang, 2007; Myers, 2005; Shiraev and Levy, 2007). There are many facets or components of culture, including nationality, ethnicity, religion, geographical area, age, hobbies, social class, and corporate culture (Cohen, 2009; Doney, Cannon, and Mullen, 1998). However, to understand this construct, researchers have often used nationality as a proxy for culture (e.g., Chinese representing Eastern cultures and North Americans representing Western cultures) and have developed dimensions of culture to evaluate their influence on various outcomes. Among the most cited cultural frameworks is Hofstede's Cultural Values, which contains five dimensions: *power distance*, *collectivism/individualism*, *uncertainty avoidance*, *masculinity/femininity*, and *long-term/short-term orientation* (Hofstede, 1980; Hofstede and McCrae, 2004). Hofstede's framework has fueled the development of many theories of culture including the Global Leadership and Organizational Behavior Effectiveness (GLOBE) project (Javidan and House, 2001), Cultural Mosaic (Chao and Moon, 2005), World Values Survey (Inglehart, 1977), and GlobeSmart Commander (Sutton and Pierce, 2003).

## The Intercultural Model of Trust

The black boxes in Figure 8.1 represent our adaptation of the Mayer et al. model. Our goal is to advance the understanding of how trust can develop or go awry when those who interact differ on cultural factors. The Intercultural Model of Trust proposes that *cultural distance* and *cultural biases* are significant antecedents to the trust-building process. Williams (2001, 2004) included trustor's *affect*, or feelings, in her model of trust; however, it was not included in the Mayer et al. model. It has been added to the adapted model because it is proposed to play a strong role when cultures are similar (helping trust) and when cultures clash (hurting trust). Finally, we included a moderator that often plays a role in stability operations where trust is needed: *conflict/instability*.

## Cultural Distance and Biases as Antecedents of Trust

We propose that before a trustor makes any judgments about *ability*, *benevolence*, or *integrity* and before s/he develops any feelings towards a potential trustee, two major variables must be considered.

First, the magnitude of cultural differences (*cultural distance*) plays a role because the greater the distance, the more difficult it is for the trustor to make accurate judgments and have positive feelings about the trustee. Second, culture-based *cognitive biases* (e.g., biases based on stereotypes) exert an influence by coloring the the potential trustor's perceptions of and judgments about trustworthiness cues that the trustee exhibits.

### *Cultural Distance*

In stability operations, personnel often work with local elders and residents, as well as foreign security forces and non-governmental agencies. Compared to US military personnel, these individuals may be extremely different in terms of language, values, norms and other variables associated with culture. The more differences between two cultures (and the greater the magnitude of these differences), the wider the cultural distance. For example, the national cultures of the US and China would be considered to be more distant than those of the US and Canada. It is logical to assume that *the wider the cultural distance*, the more difficult trust building will be. We would expect this to be the case because cultural distance has an effect on how comfortable people will be with each other (Triandis, 1994), on communication effectiveness, and predictability.

Emotional comfort is an issue because working with culturally-distant individuals will often force the interactants to cope with unfamiliar processes, environments, and expectations. In addition, people naturally have fewer positive feelings when interacting with people who are not part of their in-group (Brewer, 1979, cited in Williams, 2001). They are likely to be less comfortable and have less social contact with them, compared to people from their in-group. When working with culturally-distant people, they will be less able to rely on interpersonal similarity (i.e., common or shared backgrounds and experiences) to produce mutual attraction and a willingness to work cooperatively. Thus, the degree of cultural distance has a bearing on the trustor's feelings (*affect*), leading to either emotional discomfort or attraction.

Not only does cultural distance impact feelings, but it also impacts trust by straining communication, creating conflict and ambiguity. When people from different cultures interact, misunderstandings and conflict frequently occur (Burke, Sims, Lazzara, and Salas, 2007). Even when there is no clear conflict, communications and behavior are more ambiguous. A lack of common frames of reference, values, and behavioral expectations add to the overall uncertainty of the situation (Chao and Moon, 2005). Compared to monocultural situations, it is more difficult to understand others and it is more challenging to know if others understand you. This often produces an ambiguous state, where those from both cultures are uncertain of what is going on. Communication difficulties affect the extent to which a trustee is able to convey trustworthiness cues and the extent to which the trustor can pick up on them.

Predictability is also related to cultural distance, with people being less able to predict each other's behaviors and communication patterns. According to Adams, Bryant, and Webb (2001), trust is based on expectations about whether or not another person is likely to behave consistently and positively. If people come from the same culture, then there is a strong foundation from which they can predict one another's behavior. For example, if an American Marine is immersed in a tense tactical situation, he will have a better ability to predict the behaviors of a fellow Marine than the behaviors of an Iraqi policeman. He is able to trust the Marine because he can assume they share the same goals (relates to integrity), they have had similar training (relates to ability), and they belong to the same close-knit group (relates to benevolence).

With enough understanding of a different culture, 3C can play a strong role in smoothing interactions with a culturally-distant person. A trustee can convey trustworthiness cues in a way that is meaningful and appropriate for the trustor, thus reducing ambiguity. S/he will be able to act in a manner that the trustor will find predictable, given repeated interactions. Understanding culturally-appropriate behaviors that encourage positive feelings is another aspect of cultural competence that can accelerate trust building. People who effectively engage in cross-cultural interactions use the other culture's social customs to bridge the gap, as well as to draw attention to what they have in common. For example, many US military personnel engage in culturally-appropriate small talk with their Arab security force counterparts in the Middle East by focusing on their children (especially male children). Asking about male relatives (who have a valued status in Arab culture) elicits feelings of comfort and other positive emotions.

*Cultural Biases*

A bias is a preference toward a particular perspective or ideology that interferes with the ability to be impartial and objective (Myers, 2005). The cognitive information processing literature demonstrates that while individuals engage in observing, storing, and applying knowledge, the process is not free from bias and error. Individual perspectives and preferences influence this process to make information relevant to them. The concept of cognitive biases based on cultural norms and values was an important research focus in the NATO Human Factors and Medicine Research and Technology Group 138 (HFM RTG-138) Adaptability in Coalition Teamwork (see Sutton and Edelmann, 2005; Johnston and Mangos, 2008; Sutton and Pierce, 2003). We leveraged findings from this research to examine how culturally-based biases may affect interpersonal trust building between individuals from different cultures.

We expect that cultural biases have a powerful influence on perceptions and sense-making in social interactions, especially in ambiguous multicultural contexts where there is much room for interpretation. People will pay attention to different aspects of a social situation, value different outcomes, and make different attributions, depending on their cultural biases. These differences, in turn, will affect their behaviors. Thus, we expect cultural biases to affect the way people read, react to, and weigh cues about ability, benevolence, and integrity. We also expect cultural biases to influence their feelings about those who are trying to gain their trust.

*In-group bias* Ross (1977) coined the term *fundamental attribution error*, to explain 'the tendency for some observers to underestimate situational influences and overestimate dispositional influences upon others' behavior' (Myers, 2005, p. 88). The research on fundamental attribution error is analogous to that of in-group bias. In-group bias is the tendency for people to favor those who share a common identity (with their in-group) compared to those who are viewed as different (their out-group). When a member of an in-group exhibits a desirable behavior, the cause is attributed to positive traits. However, when a member of an in-group makes a mistake, the cause is attributed to the situation. On the other hand, if an out-group member does something positive, it would be perceived as being a result of the situation, and if s/he made a mistake, it would be perceived as being a result of disposition. Essentially, in-group members are given the benefit of the doubt, whereas out-group members are not (Brislin, 1981). Therefore, the in-group bias might make it difficult for people from different cultures to trust each other.

*Stereotypes* To avoid overload, individuals engage in stereotyping various groups of people – both in their in-groups and out-groups.. Stereotypes are the mental snapshots that people form about other groups (e.g., people from the US are aggressive). People use stereotypes to take some of the guesswork out of interactions with dissimilar individuals to reduce feelings of unpredictability (Gudykunst, 1995). Understanding others' stereotypes of US military personnel can provide insight into the appropriate cues to display when attempting to gain trust. Knowing how to accurately convey trustworthiness to people from other cultures will likely emerge as a critical 3C skill.

Research supports that individuals within the same culture share similar values, and thus, similar biases (Kahan, 2008). These values and biases are manifested in beliefs, attitudes, and behaviors. The possibilities for thinking about how culture-based biases might impact trust are nearly infinite. We will focus on two culturally-based values (and their accompanying biases) that are most likely to affect people's willingness to engage in trust and accept risk, especially in a multicultural setting: Hofstede's individualism/collectivism and uncertainty avoidance.

*Individualism/collectivism* Generally speaking, the US is considered an individualistic society, whereas most of the people with whom military personnel interact in stability operations are from collectivistic cultures. By definition, people from individualistic cultures value individual goals over group goals (Hofstede, 1980). In contrast, collectivists, by nature, tend to be more group oriented, with their identity centered around 'we'. They also make clear distinctions between in-groups and out-groups. Although they are intensely loyal to their in-group members, when dealing with out-groups, collectivists tend to interact less, share less information, and place less value on the input from out-group members (Salas, Burke, Wilson-Donnelly, and Fowlkes, 2004). In a sense, we can say that collectivists may have a pronounced in-group bias.

Additionally, these cultures differ when perceiving trustworthiness cues. Branzei et al. (2007) reported that individualists bestow trust based on a trustee's perceived ability and integrity, but collectivists' trust choices depend to a greater extent on predictable and benevolent interactions with a potential partner. Collectivists also differ from individualists in that they are more sensitive to contextual factors (Gudykunst, Ting-Toomey, and Chua, 1988).

A person from an individualistic culture relies on explicit and direct communication, emphasizes verbal over nonverbal communication, separates tasks from relationships, emphasizes individual initiative and decision making, and prefers linear reasoning (Reynolds and Valentine, 2004). On the other hand, a person from a collectivistic/high context culture relies on implicit communication, emphasizes nonverbal communication, puts relationships over task, emphasizes collective initiative and decision making, and relies on intuition and trust over facts and statistics (Reynolds and Valentine, 2004). Taken together, these differences suggest that when US military personnel are trying to gain the trust of people in collectivist cultures, they need to take a radically different approach from the way they might normally operate. While direct, straight-to-the-point communication might actually be tied to trustworthiness in their own culture, this approach is directly at odds with the values and biases of a collectivist. Instead, time and energy need to be invested to develop relationships before trust is expected, requiring a less direct and more patient communication style.

*Uncertainty avoidance* People are often faced with uncertainty, but how individuals cope with this uncertainty differs by culture. Uncertainty avoidance is defined as the extent to which the members of a culture feel threatened by ambiguous or unknown situations (Hofstede, 1980). In such societies, people are socialized to believe that avoiding uncertainty is important. One strategy commonly used is strictly conforming to established rules (Brislin, 2000). In contrast, people with lower uncertainty avoidance may dislike having too many rules that limit freedom.

Uncertainty avoidance also affects risk-taking behaviors. People who are higher in uncertainty avoidance are more hesitant in decision making and less likely to take risks (Brislin, 2000; Hofstede, 1986). We propose that people who are high in uncertainty avoidance may find unpredictable situations, such as multicultural interactions, more stressful than those who are more tolerant of uncertainty.

*Summary of cultural antecedents* In summary, all cultures are subject to bias. In-group bias is an example of a bias that most cultures share, and therefore, will always be a consideration when developing trust relationships, especially when those who interact come from different groups or cultures. In addition, values specific to certain groups/cultures such as individualism/collectivism and uncertainty avoidance are likely to affect trust-building efforts. These biases affect perceptions of trustworthiness and influence the possibility of engaging in a risk-taking relationship.

## Inputs of the Intercultural Model of Trust

*Affect* was embedded in the IMT as an additional input (with *perceptions about ability*, *benevolence*, and *integrity* being the initial inputs) to trust because it is proposed to play an important role in cross-cultural trust building. Culturally-similar individuals are more inclined to develop positive feelings than those who are culturally distant. Further, feelings that a potential trustor may have towards a trustee are likely to be affected by bias. For example, if a potential trustor holds a negative stereotype about a group to which the trustee belongs, that trustor may have negative feelings towards that person and will be less likely to trust.

The relation between affect and cognitions about trustworthiness is a complex link within the IMT. Although it is related to, and can build on cognitive-based trust, affect-based trust may be a separate construct (McAllister, 1995). Affect-based trust is enhanced when close, personal attachments lead to caring and benevolent actions, which then builds positive affect. Consequently, these feelings shape perceptions of trustworthiness (McAllister, 1995).

Cognitions about trustworthiness also impact feelings a person will have for another individual. For example, a trustor will think 'if you (the trustee) do something to harm me, I may judge that you do not have my best interests at heart, and I will update my judgments about your integrity and benevolence. I will also be likely to develop negative feelings about you.' An in-depth discussion of this relationship is beyond the scope of this paper, but one basic aspect of it is that affect, such as fear or threat, can be viewed as a source of information that influences cognitions about benevolence (Williams, 2004).

The role of affect in trust building may be particularly relevant to mixed-culture settings because people naturally have more positive feelings when interacting with others from their in-group (Brewer, 1979, as cited in Williams, 2004). Feelings can be positive, neutral, or negative, and can influence the initial levels of trust, as well as the future trajectory of that trust (Brewer and Brown, 1998; Williams, 2004). According to Williams (2004), affective states and affective attachments influence how people evaluate others' trustworthiness, how motivated they are to trust others, and how inclined they are to help others. People often rely on 'gut' feelings in deciding whether or not to trust someone. However, it is important to understand that some level of cognitive-based trust may be necessary before affect-based trust can occur (McAllister, 1995).

The role of affect is also critical for the iterative trust-building or deterioration process by being part of a feedback loop. Whether the consequences from the trustor–trustee interactions are favorable (e.g., the trustee kept his/her end of the deal) or unfavourable (e.g., the trustee reneged), cognitions and feelings would be updated for future trust decisions.

## Conflict/Instability as a Moderator of the Intercultural Model of Trust

Conflict/instability may moderate the relationship between an individual's cognitions and feelings about a trustee and his/her ability or willingness to extend trust. Inglehart and Baker (2000) presented an analysis of a series of national level surveys of attitudes, values, and beliefs around the world (covering 65 countries in six inhabited continents). Their results suggest a relationship between the levels of trust and security in a nation. The authors describe 'survival values' and 'self-expression values' at each end of a single continuum. A trend toward self-expression taps a syndrome of trust, tolerance, subjective well-being, and political activism that emerges in post-industrial societies with high levels of security. Individuals born into this cultural setting take security for granted. Without any overwhelming threat to their lives, they are free to pursue self-expression, subjective well-being, and quality-of-life concerns (Inglehart, 1977, 1990, 1997).

People born into a cohort with extreme survival values have lives shaped by insecurity and low levels of well-being, tending to emphasize economic and physical security above all other goals. They also tend to feel threatened by foreigners and other out-groups and are resistant to cultural change. Furthermore, when people are forced to live under conditions of repression and deprivation for long periods of time, their culture may become a primary source of stability and dignity (Joint Doctrine Note, 2009). These people may be hypersensitive to any slight, humiliation, or threat to their culture. Two implications follow for American service members working abroad. First, it may be extremely challenging to establish trust with individuals who are fighting for survival because they may be less likely to trust in general, and especially unlikely to trust foreigners. Secondly, the need for in-depth knowledge of how to communicate and work with culturally-distant individuals while using culturally correct, polite, and respectful practices may be critical in order to build relationships.

Additionally, people who are in the unfortunate position of being born to parents who have suffered through a legacy of instability or threat (political or otherwise) may be less likely to trust others who are not like them (not from the same family, tribe, nation, etc.) for the simple reason that cultural views and values are handed down from generation to generation (e.g., 'if my father does not trust foreigners, then I will be less likely to trust foreigners'). We would expect for an environment of instability and conflict to moderate the relationship between trust inputs (cognitions and affect) and the decision to trust. It may be difficult for a trustor to accept more vulnerability in an already insecure environment where there are numerous unknowns.

The IMT, the following scenario provides an operational example of how cultural distance, biases, affect, and environmental factors influence trust.

## Scenario

> Sgt. Christopher Johnson had recently been deployed to Darfur, Sudan – the center of 21st century's first genocide – to support a US Army squad with the relief missions dedicated to assisting the people displaced to the Kolbus refugee camp. His primary objective was to build trust with the tribe and to get cooperation from the local men. He needed drivers to transport the food and supplies to the camp. However, the Janjaweed had attacked the drivers the day before, and this prevented the men from volunteering. The Janjaweed, supported by some Sudanese government officials, are responsible for the persistent violence and instability within the region. Sgt. Johnson decided to meet with the tribal leader, Ibrahim, to discuss the previous day's attacks and plans for future deliveries.
>
> Carrying a small crocus bag of salt, Sgt. Johnson approached the leader's compound after their morning prayer. He offered the salt and crossed his heart while bowing his head and said, 'Asalam Alakum.' Ibrahim responded, 'Wa alaikum asalaam.' He acknowledged the Sergeant's sign of respect and benevolence, and graciously welcomed him inside the circle among the crowd of men. Sgt. Johnson's voice trembled as he conveyed his concern for the

> kidnapping of five of the tribe's men. After a period of silence, Ibrahim nodded for the outsider to continue. Sgt. Johnson explained that the Janjaweed were responsible for the incident and that the military brought additional troops to rescue them. He explained the rescue strategy. He continued by stating that he understood that the drought created a need for additional clean water and food for the tribe. Since their supplies were not delivered, he needed additional volunteers (who knew the back roads) to accompany the military to obtain the awaiting supplies. After he conveyed the dilemma, he assured that there would be additional security to reduce the risk of being captured by the Janjaweed.
>
> Ibrahim noted his sincerity and asked Sgt. Johnson to step out as he discussed the request with his men. After a long period of careful deliberation, Ibrahim brought out Souleymane, Mohammed, and Salim to join Sgt. Johnson for mission planning.

*Reflection about Sgt. Johnson* During this scenario, we illustrated how a cross-culturally competent military leader should behave to engender trust. Ibrahim's tribe was in the midst of a horrific (unstable/conflict) environment, with some of his people having been recently kidnapped. Sgt. Johnson was asking for cooperation that entailed risk (thus, required trust). He needed for Ibrahim to provide additional tribesmen for a dangerous mission with the US military team at the helm.

Sgt. Johnson made every effort to bridge the cultural distance between his own US culture and Ibrahim's tribal Sudanese culture by respectfully greeting them in their native Arabic language and using tribal gestures. He also presented a valued gift and took time to build rapport before redirecting the conversation to his request. These behaviors enhanced Ibrahim's positive feelings toward Sgt. Johnson (affect), making him feel comfortable and allowing him to see that the Sergeant had his best interests in mind (benevolence).

Sgt. Johnson overcame his own biases and adapted to those of his Sudanese counterparts by acknowledging that he was a member of the out-group; thus, he changed his stereotypical individualist communication patterns. He recognized the importance of showing concern regarding the kidnapping of the tribal men, rather than rushing to the point. He also provided information about how the mission has now changed to include additional security, which conveyed ability and benevolence. Also, explaining the new strategy to Ibrahim helped reduce the level of uncertainty and perceived risk for the volunteers. By demonstrating effective cross-cultural behaviors (and genuine concern), Sgt. Johnson was able to accomplish the mission by recruiting more drivers.

## Implications

### *Practical Implications*

In the following sections, we describe the practical implications of what has been discussed thus far, including the value of cross-cultural competence (3C) in trust building, effective cultural behaviors, implications for using interpreters, and finally, managing affect using threat reduction.

*Cross-cultural competence* It was the primary objective of this chapter to propose and discuss cultural variables that affect trust. To merge our theory with the operational environment, we first discuss the practical implications of 3C—a multidimensional construct that develops over time, promotes clearer communication, builds trust, and strengthens multicultural relationships (McCoy and Severe, 2010). Trustees that possess 3C may be more successful in developing relationships because they are more aware of and open to interacting with other cultures. Training for 3C may be useful in better facilitating trust-building activities in multicultural environments.

A major component of 3C is being aware of one's own culture and ways that it might clash with the culture of others. Military personnel need to be aware of those aspects of 'cultural baggage' that are likely to irritate or offend people in host-countries. Simply put, we are not likely to build trusting relationships with those whom we offend or irritate. Military personnel who are exposed to 3C training, and are able to move beyond ethnocentrism, will begin to gain an awareness of the need to adapt to those from other cultures (e.g., where rules and laws may be governed by tribes and religion). Table 8.1 gives common examples of how Western values may be perceived differently by other cultures and how to adapt behaviors to overcome these challenges..

A second major aspect of 3C is understanding the culture of others. Although culture has been defined in many ways, many have suggested that a person's 'cultural lens' will color his/her perception of the world around him/her. Klein, Pongonis, and Klein (2000) have stated that multinational team members could improve the team's performance if they could see the world though one another's eyes. Indeed perspective taking has been defined by multiple researchers as a key 3C capability (Epley, Keysar, Van Boven, and Gilovich, 2004; Galinsky and Moskowitz, 2000, cited in Abbe et al., 2007).

In order to be able to take the perspective of another person, we need to have knowledge about that person and their culture. Perspective taking is an interpersonal skill, which may develop over time and with experience.

**Table 8.1 Interpretations of Western idioms and recommendations for cross-cultural behaviors**

| Dimensions | Western idioms | Other possible interpretations | Recommended behaviors for individualists operating in collectivist cultures |
|---|---|---|---|
| Individualism | 'If you want a job done right, do it yourself' | Self-centered and poor team players | • Recognize and show respect for the group identity |
| Task orientation | 'Keep your eye on the ball' | Unfriendly and lack of caring | • Respect the value and need for relationship building in other cultures |
| Self-reliance | 'Pull yourself up by your own boot straps' | Turning your back on those in need | • Recognize the value of interdependence in other cultures |
| Egalitarianism | 'All men are created equal' | Unwilling to show proper deference and respect to those who deserve it | • Show proper deference and respect to those in power in the host culture |
| Time consciousness | 'Time is money' | Impatient, rushed, and pushy | • Be patient and allow more time than you expect for meetings<br>• Do not view tardiness to meetings by members of a host culture as a sign of disrespect |
| Content focus | 'Get to the point' | Naive or narrow-minded | • Display patience and important nonverbal cues and other communication context variables relevant to the host culture<br>• Use silence to enhance comfort level in face-to-face communication |
| Future orientation | 'Don't cry over spilled milk' | Ignorant of past injustices | • Be knowledgeable about past interactions between the two cultures<br>• Avoid perpetuating negative stereotypes |
| Directness | 'Don't beat around the bush' | Blunt to the point of rudeness, lack of subtlety, and sophistication | • Be less blunt and direct<br>• Avoid direct questions that might call for accountability, especially if it might seem like blame<br>• Avoid drawing attention to mistakes made by members of the other culture<br>• Make decisions consensually and leave time for discussing particulars |

*Cultural behaviors* Abiding by another person's cultural social conventions or norms is helpful or even necessary to develop a warm, friendly relationship. As Joint Doctrine Note (2009) points out, social conventions such as how to enter a home, who to talk to, how to conduct meetings, the acceptance of hospitality, and the giving of gifts shape everyday interaction and are a reflection of culture. Some cultures place a considerable amount of importance on such traditions, social rules, formality, and status roles. Indeed, when there is a strict code of conduct in a culture, individuals whose trust we need to earn may be easily offended by our lack of adherence to this code. This suggests why the most conspicuous aspect of culture training is often the do's and don'ts for operating within a culture.

*Trust building when using interpreters* In the military, it is very common to employ interpreters to communicate with individuals from different countries. The use of interpreters becomes increasingly necessary when the mission objectives are time-sensitive and the consequences of a misstep are dire. Although trust is often developed over time, there are some instances where a level of trust is needed immediately. Therefore, engaging in swift trust with an interpreter is often required to accomplish mission-critical tasks. Humanitarian military missions that require working with local civilians and government officials, different military branches, foreign military services, and non-government organizations, are a prime example of when swift trust is necessary for mission effectiveness. Moreover, having a trusting relationship between the trustee (US service member) and the interpreter prior to trying to build one with the trustor is recommended.

Military personnel relying on interpreters must be confident that their interpreters are willing and able to manage trustworthiness cues for them so that they appear trustworthy. The interpreter literature offers some practical suggestions for effectively matching interpreters with trustees (i.e., considering gender, age, etc.; Wallin and Ahlström, 2006). Overall, an exceptional interpreter is one who is competent about the topic being discussed, as well as both languages and cultures (Murray and Wynne, 2001). More importantly for intercultural communications in stability operations, the interpreter should not only be cross-culturally competent, but also have knowledge about both military (e.g., rank, mission, expectations) and civilian norms (e.g., status, local leaders, traditions) from both cultures.

A critical issue with working with interpreters is not being fluent in the host culture can jeopardize the trust-building relationship by simply translating words and not considering the cues and context that surround the interaction. For example, a US military person might display an acceptable level of benevolence and integrity, but focus on emphasizing their own ability (in accordance with American trust values). Perhaps the trustor values benevolence more than the other trustworthiness indicators. An optimal situation would be for the interpreter to, through words and gestures, display more benevolence to convey the trustworthiness intended by the trustee. *The result is that the level of trust intended and offered is the level of trust perceived.* To summarize, the interpreter's expertise in the subject matter, level of

3C, and ability to accurately translate trust intentions are all notable characteristics that can facilitate the fostering of trust between cultures.

*Threat reduction* One way that we may manage the affect of individuals from other cultures is to try to engage in threat-reducing behaviors. In any case where there is a strong sense of 'we' versus 'them,' this is probably critical. Williams (2004) described how trust and cooperation can be enhanced by reducing threat when team members from different organizations must collaborate. According to her work, threat-reducing behaviors may include: altering the situation, deflecting attention from the emotion-provoking elements, reframing the situation to seem less harmful, and directly modulating counterparts' emotional responses to the situation. This may sound difficult to translate into working across cultures, but the idea of reducing threat is quite relevant. At a minimum, we should avoid shaming or blaming those whose trust we are trying to cultivate. In some cultures (e.g., in the Middle East), respect is critical and is related to the concepts of honor and shame (Joint Doctrine Note, 2009). If honor is lost and shame results, this can lead to extremely negative feelings, and even a need for revenge. Thus, we need to be very mindful of preserving dignity in any way possible.

### *Research Implications*

The IMT in Figure 8.1 presents the opportunity to advance the research literature on trust in multicultural environments. First, we encourage validation of both the model components and relationships among the variables. The next section presents questions we believe would be fruitful avenues of research.

*What are the most effective ways for military personnel to win the trust of people who a culturally distant?* General 3C training may be helpful to begin understanding how other cultures differ from one's own, and how one's behavior could be misinterpreted. More advanced knowledge and experience with a particular culture may be needed to fully understand how to earn the trust by accurately conveying trustworthiness cues that are meaningful to individuals from that culture. Table 8.1 presents stereotypical Western concepts regarding working relationships. Our shared culture allows us to take shortcuts through the trust process. However, when we are immersed in another culture and trying to win trust, these shortcuts no longer function. For example, we win the trust of each other in our culture by getting down to business right away ('keep your eye on the ball'), being direct, and task-focused (focusing on ability). However, this mindset may not work in other parts of the world where getting down to business comes after the more important relationship building phase (assuring benevolence and integrity).

*How do cultural-specific models of trust converge and diverge?* If we take a substantial step back from our assumptions, we can begin to question whether or not the very process of trust building might be different in other cultures. We suggest that research is needed to test our basic assumptions. Thus far, our analysis of trust has had a Western bias, mainly because it reflects our culture and the literature that is readily available to us. If resources allowed, it would be ideal to gather data from a more diverse audience (e.g., examining international models of trust).

*What trustworthiness signs are valued in different cultures? Is trustworthiness cued differently?* We have been forced to make some assumptions in order to discuss the influence of culture on trust in a meaningful way. For instance, we have assumed that across cultures, trustors probably make judgments about a trustee's ability, benevolence, and integrity. Also, we assume that feelings weigh into trust decisions across cultures. However, we are uncertain about the relative weight of these inputs. The appropriate trustworthiness signs are implicitly understood by individuals from the same culture. However, when a multicultural context is introduced, the trustor may not value the cues that are being displayed by the trustee, even though they are effective in the trustee's culture. More culture-specific research is needed.

*How can we overcome in-group and other biases that are likely to shape potential trust relationships in multicultural contexts?* Being seen as an outsider can make building trusting relationships difficult. Some potential methods for reducing in-group bias include creating shared experiences, agreeing on shared goals, and using an interpreter as a liaison between culturally distant groups. These and other methods should be explored.

*How can trust be repaired among culturally distant parties?* In interpersonal relationships when there is broken trust, the violator is expected to accept responsibility, show humility, apologize, and make reparations. Importantly, the victim must see the value in continuing to expend effort and take risks in the relationship. Similar processes may be needed to repair trust when there is history between cultures that led to negative stereotypes. Showing humility and accepting responsibility for past wrongs are not normally publicly emphasized in military strategies. Therefore, the military needs to determine the right balance for displaying strength and humility, while accomplishing mission goals.

*What are the implications of the distrust/trust distinction?* The construct of distrust, and its relationship with trust, has emerged as the focus of contemporary research (Huang and Dastmalchian, 2006; Lewicki, McAllister, and Bies, 1998; Luhmann, 1979). The initial view of trust and distrust was that they were mutually exclusive – opposite ends of a single continuum. However, they are now considered two separate, yet related, constructs (Lewicki et al., 1998). The main support for the distinction is that it is possible that the same interactants can both trust and distrust each other at the same time. *Simply stated, the opposite of trust is not distrust.* 'Trust and distrust both entail certain expectations, but whereas trust expectations anticipate beneficial conduct from others, distrust expectations anticipate injurious conduct' (p. 444). Table 8.2 details the levels of trust and distrust and their combined characteristics as proposed by Lewicki et al., (1998).

Moreover, the authors proposed four interactions that are composed of both high and low levels of trust and distrust. For example, Lewicki et al. (1998) proposed that the high trust/high distrust combination is characterized by: trust but verify; relationships highly segmented and bounded; opportunities pursued and down-side risks/vulnerabilities continually monitored. Although the authors have proposed the various relationships among trust and distrust, few studies have validated them. Understanding distrust will allow us to consider avenues for rebuilding relationships when it exists.

**Table 8.2 Integrating trust and distrust: alternative social realities (adapted from Lewicki, McAllister, and Bies, 1998, p. 445)**

| | | |
|---|---|---|
| **High trust**<br>Characterized by<br>Hope<br>Faith<br>Confidence<br>Assurance<br>Initiative | High-value congruence<br>Interdependence promoted<br>Opportunities pursued<br>New initiatives | Trust but verify<br>Relationships highly segmented and bounded<br>Opportunities pursued and down-side risks/vulnerabilities continually monitored |
| **Low trust**<br>Characterized by<br>No hope<br>No faith<br>No confidence<br>Passivity<br>Hesitance | Casual acquaintances<br>Limited interdependence<br>Bounded, arms-length transactions<br>Professional courtesy | Undesirable eventualities expected and feared<br>Harmful motives assumed<br>Interdependence managed<br>Preemption; best offense is a good defense<br>Paranoia |
| | **Low distrust**<br>Characterized by<br>No fear<br>Absence of skepticism<br>Absence of cynicism<br>Low monitoring<br>No vigilance | **High distrust**<br>Characterized by<br>Fear<br>Skepticism<br>Cynicism<br>Wariness and watchfulness<br>Vigilance |

## Conclusion

The current and future military operations require a reliance on relationship building to perform stability operations. Mission success is no longer based on kinetic forces alone, but also on building trust for negotiating with and gaining cooperating from the local populace. A critical defense will be to employ the military with the necessary cross-cultural competence, along with regional and language proficiency to accelerate the mission objectives. Embracing missions with an openness for learning about the host cultures, maintaining self-awareness, recognizing others' cultural biases, and possessing the ability to adapt to that culture are valuable skills for military personnel before deployment. Without these skills, the stability operations will be jeopardized, resulting in compromised missions and possible lives lost.

Identifying factors that enhance and deter trust is critical for understanding how to develop the relationships needed for rebuilding nations and supporting humanitarian missions. In this chapter, the authors adapted the Mayer et al. model to illustrate how cultural biases and cultural distance can influence the fostering of trust in a mixed-culture setting. The Intercultural Model of Trust presents an opportunity for an initial research dialogue about trust in a cross-cultural context and drives possible training strategies for US military personnel operating abroad.

## References

Abbe, A., Gulick, L. M. V., and Herman, J. L. (2007). *Cross-cultural competence in Army leaders: A conceptual and empirical foundation* (ARI Study Report 2008–01). Arlington, VA: US Army Research Institute for the Behavioral and Social Sciences.

Adams, B., Bryant, D., and Webb, R. (2001). *Trust in teams literature review* (Report No. CR-2001–042). Ottawa, Canada: Department of National Defence.

Branzei, O., Vertinsky, I., and Camp, R. D. (2007). Culture-contingent signs of trust in emergent relationships. *Organizational Behavior and Human Decision Processes*, *104*, 61–82.

Brewer, M.B. (1979). The psychology of intergroup attitudes and behavior. *Annual Review of Psychology*, 36: 219-243.

Brislin, R. W. (1981). *Cross-cultural encounters: Face-to-face interaction.* Elmsford, NY: Pergamon.

Brislin, R. W. (2000). *Understanding culture's influence on behavior* (2nd edn). Fort Worth, TX: Harcourt.

Brewer, M. B. and Brown, R. J. (1998). Intergroup relations. In D. T. Gilbert and S. T. Fiske (eds). *The handbook of social psychology*, (4th edn, pp. 554–594). New York: McGraw–Hill.

Burke, S., Sims, D., Lazzara, E., and Salas, E. (2007). Trust in leadership: A multi-level review and integration. *The Leadership Quarterly*, *18*, 606–632.

DOD Directive 1322.18 (2009, January). Military Training. Washington, DC: Deputy Secretary of Defense.

Chao, G. T. and Moon, H. (2005). The cultural mosaic: A metatheory for understanding the complexity of culture. *Journal of Applied Psychology*, *90*, 1128–1140.

Cohen, A. B. (2009). Many forms of culture. *American Psychologist*, *64*(3), 194–204.

Department of the Army. (2008). *FM 3–07: Stability Operations*. Retrieved December 08, 2008, from: http://www.fas.org/irp/doddir/army/fm3–07.pdf.

Doney, P. M., Cannon, J. P., and Mullen, M. R. (1998). Understanding the influence of national culture on the development of trust. *Academy of Management Review*, *23*, 601–620.

Epley, N., Keysar, B., Van Boven, L., and Gilovich, T. (2004). Perspective taking as egocentric anchoring and adjustment. *Journal of Personality and Social Psychology*, *87*, 327–339.

Fisher, R. and S. Brown. (1988). Getting together: Building relationships as we negotiate. London: Penguin.

Galinsky, A. D. and Moskowitz, G. B. (2000). Perspective taking: Decreasing stereotype expression, stereotype accessibility and in-group favoritism. *Journal of Personality and Social Psychology*, *78*, 708–724.

Gudykunst, W. B., Ting-Toomey, S., and Chua, E. (1988). *Culture and interpersonal communication.* Beverly Hills, CA: Sage.

Gudykunst, W. B. (1995). Anxiety/uncertainty management (AUM) theory. In R. L. Wiseman (ed.), *Intercultural communication theory* (pp. 8–58). Thousand Oaks, CA: Sage.

Hardison, C. M., Sims, C. S., Ali, F., Villamizar, A., Mundell, B., and Howe, P. (2009). *Cross cultural skills for deployed Air Force personnel*. RAND Project Air Force. Retrieved March 3, 2011 from http://www.rand.org/pubs/monographs/MG811/.

Hofstede, G. (1980). *Culture's consequences: International differences in work-related values.* Newbury Park, CA: Sage.

Hofstede, G. (1986). Cultural differences in teaching and learning. *International Journal of Intercultural Relations*, *10*, 301–320.

Hofstede, G. and McCrae, R. R. (2004). Personality and culture revisited: Linking traits and dimensions of culture. *Cross-Cultural Research*, *38*, 52–88.

Huang, H. J. and Dastmalchian, A. (2006). Implications of trust and distrust for organizations: Role of customer orientation in a four-nation study. *Personnel Review*, *35*, 361–377.

Inglehart, R. (1977). *The silent revolution: Changing values and political styles in advanced industrial society.* Princeton, NJ: Princeton University Press.

Inglehart, R. (1990). *Culture shift in advanced industrial society*. Princeton, NJ: Princeton University Press.

Inglehart, R. (1997). *Modernization and postmodernization: Cultural, economic and political change in 43 societies*. Princeton, NJ: Princeton University Press.

Inglehart, R and Baker, W. E. (2000). Modernization, cultural change, and the persistence of traditional values. *American Sociological Review*, *65*, 19–51.

Javidan, M. and House, R. J. (2001). Cultural acumen for the global manager: Lessons from project GLOBE. *Organizational Dynamics*, *29*, 289–305.

Johnston, J.H. and Mangos, P. (2008). *Measuring cultural cognitive biases in multi-national research.* Proceedings of the North Atlantic Treaty Organization Research and Technology Organization: Human Factors and Medicine Symposium on Adaptability in Coalition Teamwork, Copenhagen, Denmark. Retrieved September 9, 2010 from http://www.rta.nato.int/pubs/rdp.asp?RDP=RTO-MP-HFM-142.

Joint Doctrine Note (2009). *The significance of culture to the military*. Joint Doctrine Publication. Shrivenham: The Development, Concepts and Doctrine Centre, Ministry of Defence.

Kahan, D. M. (2008). *Cultural cognition as a conception of the cultural theory of risk.* Cultural Cognition Project Working Paper No. 73. Retrieved September 9, 2010 from http://ssrn.com/abstract=1123807.

Klein, H. A., Pongonis, A., and Klein, G. (2000). *Cultural barriers to multinational C2 decision making.* Proceedings of the 2000 Command and Control Research and Technology Symposium (CD-ROM), Monterey, CA.

Lewicki, R. J., McAllister, D. J., and Bies, R. J. (1998). Trust and distrust: New relationships and realities. *Academy of Management Review*, *23*, 438–458.

Lewis, J. D. and Weingert, A. J. (1985). Social atomism, holism, and trust. *The Sociological Quarterly*, *26*, 455–471.

Luhmann, N. (1979). *Trust and power*. Wiley: Chichester, UK.

Matsumoto, D. and Juang, L. (2007). *Culture and psychology* (4th edn). Belmont, CA: Wadsworth/Cengage Learning.

Mayer, R., Davis, J., and Schoorman, D. (1995). An integrative model of organizational trust. *The Academy of Management Review* *20*(3), 709–734.

McAllister, D. J. (1995). Affect-and cognition-based trust as foundations for interpersonal cooperation in organizations. *Academy of Management Journal*, *38*, 24–59.

McCoy, C. E. E. and Severe, G. (2010). *Cultural interactions: An introduction to cross-cultural competence*. Unpublished manuscript. Naval Air Warfare Center Training Systems Division: Orlando, FL.

McDonald, D., McGuire, G., Johnston, J., Selmeski, B., and Abbe, A. (2008). *Developing and managing cross-cultural competence within the Department of Defense: Recommendations for learning and assessment*. Defense Regional and Cultural Capabilities Assessment Working Group (RACCA WG) Report prepared for Defense Language Office (DLO). Retrieved September 2, 2010 from http://www.deomi.org/CulturalReadiness/documents/RACCA_WG_SG2_Workshop_Report.pdf.

Murray, C. D. and Wynne, J. (2001). Researching community, work and family with an interpreter. *Community, Work and Family*, *4*(2), 151–171.

Myers, D. G. (2005). *Social psychology*. McGraw-Hill: New York, NY.

North Atlantic Treaty Organization (2008). *Multinational military operations and intercultural factors*. NATO Research and Technology Organisation, (RTO) Technical Report. RTO-TR-HFM-120. Retrieved April 14, 2011 from http://www.dtic.mil/cgi-bin/GetTRDoc?Location=U2&doc=GetTRDoc.pdf&AD=ADA496055

Reynolds, S. and Valentine, D. (2004). *Guide to cross-cultural communication*. Upper Saddle River, NJ: Prentice Hall.

Ross, L. (1977). The intuitive psychologist and his shortcomings: Distortions in the attribution process. In L. Berkowitz (ed.), *Advances in experimental social psychology* (Vol. 10). New York: Academic Press.

Rousseau, D., Sitkin, S., Burt, R., and Camerer, C. (1998). Not so different after all: A cross-discipline view of trust. *Academy of Management Review 23*(3), 393–404.

Salas, E., Burke, C. S., Wilson-Donnelly, K. A., and Fowlkes, J. E., (2004). Promoting effective leadership within multicultural teams: An event-based approach. In D. Day, S. J. Zaccaro, and S. M. Halpin (eds), *Leader development for transforming organizations* (pp. 293–323). Mahwah, NJ: LEA.

Shiraev, E. and Levy, D. A. (2007). *Cross-cultural psychology: Critical thinking and contemporary applications* (3rd edn). Boston, MA: Allyn and Bacon.

Sutton, J. L. and Pierce, L. G. (2003). *A framework for understanding cultural diversity in cognition and teamwork*. Proceedings of the Eighth International Command and Control Research and Technology Symposium. Retrieved March 3, 2011 from http://www.dodccrp.org/ 8thICCRTS/Pres_track1.htm.

Sutton, J. L. and Edelmann, V. (2005). *Leader and Team Adaptability in Multinational Coalitions (LTAMC): An international research project*. Proceedings of the Tenth International Command and Control Research and Technology Symposium. Retrieved March 3, 2011 from http://www.dodccrp.org/ 10thICCRTS/Pres_track9.htm.

Ting-Toomey, S. and Oetzel, J.G. (2001). *Managing intercultural conflict effectively*. Thousand Oaks, CA: Sage Publications Inc.

Triandis, H. C. (1994). *Culture and social behavior*. New York: McGraw-Hill.

Tzafrir, S. S. (2005). The relationship between trust, HRM practices and firm performance. *International Journal of Human Resource Management*, *16*, 1600–1622.

Wallin, A. M. and Ahlström, G. (2006). Cross-cultural interview studies using interpreters: systematic literature review. *Journal of Advance Nursing*, *55*(6), 723–35.

Williams, M. (2001). In whom we trust: Group membership as an effective context for trust development. *Academy of Management Review*, *26*, 377–296.

Williams, M. (2004, August). *Building genuine trust strategically: A threat regulation model of cooperation across boundaries*. Paper presented at the annual meeting of the American Sociological Association, San Francisco, CA. Retrieved September 9, 2010 from http://www.allacademic.com/meta/p110896_index.html.

# SECTION THREE
# Trust in Technology

# Chapter 9
# A Reliance Model for Automated Combat Identification Systems: Implications for Trust in Automation

Justin G. Hollands
*Defence R & D Canada, Toronto, Canada*

Heather Neyedli
*University of Toronto, Toronto, Canada*

## Introduction

Automation is becoming a more prevalent component of human-technology systems, from everyday technology like cars, computers, and cell phones, to highly specialized systems like nuclear power plants and advanced aircraft. Not surprisingly, this trend is also occurring in military systems, in land, sea, and air environments. Such technology is even being extended to the dismounted soldier's kit, with prototype combat identification (Combat ID, or CID) systems now being fielded (Sherman, 2000; Sherman, 2002).

CID systems are typically based on a radio-frequency identification (RFID) technology, with the RFID tag built into the soldier's helmet or uniform, or on a vehicle platform. When a soldier targets a friendly soldier or vehicle, the RFID tag is detected, and a light or display indicates the presence of a 'friendly.' The reliability of this 'friendly' feedback is extremely high. However, 'unknown' feedback (i.e., when no RFID tag is detected) can be ambiguous, because it can represent a number of situations with different implications for action. The target could be an enemy; however, the target could be a friendly soldier whose RFID tag is not detected. This could occur for a number of reasons, such as the range of the target, problem with the sensor, weak signal strength, or loss of power. Thus, one must consider the 'unknown' feedback from a CID system to be less than perfectly reliable.

Note that this assumes that the only targets present in the battle space are friendly or hostile. However, matters are much more complex operationally. Today's soldiers work in joint coalitions, whose constituent partners may not use compatible technologies. It is also likely that neutral civilians, local police, security personnel, or members of various NGOs (non-government organizations like Médecins sans Frontières) will be present in the battle space. It is, of course, unrealistic to expect an automated CID system to be able to distinguish among all possible identity classes. Nonetheless, the soldier will still see 'unknown' feedback in these cases. Thus, the soldier is faced with the task of distinguishing these individuals from hostile forces while an automated system shows 'unknown' feedback.

*Team Trust and CID*

Trust in automated systems shares many characteristics with human trust in teams or interpersonal relationships. Indeed, some team researchers have begun to think of the human–automated system combination as a small group or team in which one member is not human (Bowers, Oser, Salas, and Cannon-Bowers, 1996; Scerbo, 1996). As effective CID requires cooperation and interaction between various individuals and groups, the implications of trust on CID are broadly considered, followed by a discussion of trust in an automated team member.

Military teams require a high level of trust, as many individuals in the team are trained for a specialized role within the working group. Individual members must rely on the performance of others within the team to achieve their individual goals, and the team's common goals. Increased trust indirectly affects performance of a team through improved communication and a reduction of conflict among team members (Dirks and Ferrin, 2001). In a diverse team, there may be a difference between the values and goals of individual members in relation to the team (Adams and Webb, 2003). In addition, there may be a lower level of perceived similarity among team members (Webber, 2002), reducing the affective and analogical elements of trust. A diverse team may therefore challenge the formation and maintenance of trust, and adversely affect team performance. Effective CID by a unit requires a high level of situational awareness (SA). Trust among team members should enhance communication and lead to improved SA.

As described above, multinational organizations with diverse doctrines, standard operating procedures and roles are operating in the battle space, which requires team cooperation at the organizational level. For instance, NGOs are unlikely to have standardized CID procedures as they are not in a combat role; however, knowledge of their location and operations can assist in battle space SA. A lack of inter- and intra-organization cooperation and communication was demonstrated by two coalition airstrikes flown against the International Committee of the Red Cross (ICRC) in Kabul twice within 10 days in 2001 (Young, 2005). Initially the ICRC failed to include the Kabul compound on a list of facilities sent to the coalition chain of command, which resulted in an airstrike on 16 October. Although location information was forwarded to Coalition headquarters after the incident, this information did not reach the appropriate individuals and the same facility suffered another airstrike on 26 October.

Inter-organizational challenges in modern combat situations have led the Canadian Department of National Defence to call for a JIMP (Joint, Interagency, Multinational, Public) approach (Gizewski and Rostek, 2007). JIMP acknowledges the role of different team members and stresses communication and sharing of resources to integrate efforts in the operational theatre. An integrated approach should lead to greater predictability, openness, and accessibility, the process bases of trust (Lee and See, 2004). A JIMP culture that fosters trust among organizations should improve information sharing and shared SA.

Effective CID is characterized by a lack of error rather than a specific output or salient goal being reached. Conversely, errors are salient given their severe consequences. Considering that ability and performance provides an important basis for trust, errors can quickly erode trust among organizational members (Mayer, Davis, and Schoorman, 1995).

### *Trust in Automation*

To help improve CID, technical solutions like RFID automation have been introduced. Automation can provide many benefits; however, the costs of introducing automation have been well documented (e.g., Bainbridge, 1983). Users must rely on the automation appropriately to gain the benefits afforded by the automation. When users over-rely on an imperfect system it is called *misuse*; *disuse* occurs when users reject the automation's capabilities (Parasuraman, Molloy, and Singh, 1993). Trust can mediate reliance behavior. Just as with human colleagues, the user should develop an appropriate level of trust in an automated partner, in order to rely on the automation appropriately (Lee and See, 2004).

Human–automation trust develops similarly to the interpersonal and inter-organizational trust discussed above. However, there are some important differences in the way humans trust automated systems. First, while naïve operators believe that they would outperform a human partner, they are also more inclined to expect their automated partner to outperform them (Dzindolet, Pierce, Beck, and Dawe, 2002). Second, users are more willing to trust a human expert than a machine (Jian, Bisantz, and Drury, 2000). Third, automation lacks intentionality and therefore purpose-based characteristics like loyalty and benevolence. Operators will attribute a similar level of knowledge but not effort to an automated system (Lerch, Prietula, and Kulik, 1997). To foster trust, the user should have a clear understanding of the purpose of the automation (Rasmussen, Pejtersen, and Goodstein, 1994). Fourth, one interacts with automation solely to achieve a goal rather than to build an interpersonal relationship; therefore, performance becomes a more important basis for trust (Madhavan and Wiegmann, 2007). Important to the assessment of performance is a bias where people initially expect machines to perform at near-perfect rates (Dzindolet, Pierce, Beck, Dawe, and Anderson, 2001b). Errors become salient because they violate the user's expectations, which results in a decline of trust due to a perceived decline in performance (Wiegmann, Rich, and Zhang, 2001).

The differences between human–automation and interpersonal trust have two implications when considering a model of human–automation performance. First, the biases and expectations regarding automation performance generally indicate that operators initially expect a high level of performance from the automation and that trust in the automation technology is fragile. A description of human–automation interaction must account for these biases and beliefs. Second, automation performance greatly affects the operator relationship with the automation; therefore a model of human performance and trust using automated systems must reflect the reliability of the automation.

In a seminal review, Lee and See (2004) propose a model in which they posit that trust is an attitude based on beliefs about automation. The level of trust – along with factors like the user's workload and the effort required to engage the automation – influences the user's intentions and ultimately reliance behavior. Consequently, trust must be calibrated to the automation's capabilities to produce appropriate reliance. Resolution and specificity also define the relationship between human trust and automation capability. Good resolution implies that a change in the capability of the automated system produces a corresponding change in trust (see also Cohen, Parasuraman, and Freeman, 1998). Lee and See argued that specificity has functional and temporal components. Functional specificity implies that trust should reflect the differing capabilities of automated system's functions and modes. Temporal specificity is the degree to which trust reflects fluctuations in capability due to time or changing situation. The model is dynamic such that automation influences trust, and trust influences interaction with the automation. This structure implies that automation must be used (i.e., relied on) for trust to develop (Muir and Moray, 1996). Whether a human operator trusts an automated decision aid will also likely depend upon the operator's self-trust or self-confidence: that is, how the operator perceives the reliability of manual operation.

Research on trust and automation has traditionally focused on supervisory control systems in domains such as nuclear energy and process control. In these systems the user's role is to monitor the automation and detect automation failures. A common problem in such systems is that operators become out of the loop with respect to the automated processes (e.g., Endsley and Kiris, 1995). These systems contrast with decision aiding automation found in CID, where the automation assists with a single task in which the user is actively engaged (Wang, Jamieson, and Hollands, 2009). The automation can be considered a cognitive tool where the user works directly on the environment with the automation assisting their interactions (Lee, 2006). In supervisory control the automation can be better described as a prosthesis which replaces a human function, or an agent which acts as a semi-autonomous partner. In the case of agents, the operator may not understand the agent's processes, reducing predictability, and trust in turn. This contrasts with CID automation which fails in a predictable way (i.e., when 'unknown' feedback is received), making specificity of trust especially important in this context. Additionally, CID is performed in a physically and cognitively demanding environment where personal risk is high, which will likely increase arousal (Wang et al., 2009).

Trust in an automated system does not occur in isolation. The operator is likely part of a team and larger organization, which influences the operator's attitude toward the automation. This influence can be formal (e.g., an organization mandates that automation be used) or informal (e.g., gossip among members of a team). Organizations have a cultural context that can influence and define the role and relationship with automation (Baba, Falkenburg and Hill, 1996). These factors are especially important if an individual has not yet interacted with the automation because they set initial beliefs about the system (Kramer, 1999; Lee and See, 2004). Thus, teams and organizations set the context for automation use which affects the soldier's belief about the properties of the automation. A model of trust in CID must reflect this belief setting.

*Belief, Trust, and Reliance*

Lee and See (2004) argued that 'trust stands between beliefs about the characteristics of the automation and the intention to rely on the automation' (p. 50). In other words, trust mediates the relationship between belief in the reliability of the automated system and human reliance on that automated system. This idea is captured in their conceptual model of the dynamic process of trust formation. Beliefs affect attitudes (such as trust), which affect reliance behaviors in turn (Ajzen and Fishbein, 1980). Beliefs are longstanding and reflect contextual variables like gossip or reputation, and in automation, interface features. Attitudes represent affective evaluation of beliefs that can guide people to action and are more situation specific (Lee and See, 2004). Trust, as an attitude, is affected by variables like organizational structure, culture, and an individual's predisposition to trust.

What are the implications of the Lee and See (2004) model for a soldier using the CID decision aid with imperfect 'unknown' feedback reliability? The soldier adjusts reliance on the system based on a belief about the CID system's failure rate. In a study using a first-person shooter game engine to simulate a combat environment, Wang et al. (2009) showed that belief in the 'unknown' feedback from the simulated CID system correlated with subjective ratings of trust in that feedback. Trust in the feedback was also correlated with reliance on the CID system. However, belief did not correlate with reliance directly, suggesting that trust plays an important intermediary role. Specifically, belief in the 'unknown' feedback affects trust in the feedback, which in turn affects how the reliance on that feedback is set. This result is consistent with the Lee and See model.

The overall relationship between belief, trust, and reliance described above only paints part of the picture. Wang et al. (2009) also manipulated whether the system's reliability level (67 percent or 80 percent reliable) was disclosed to participants. Their results showed that if the CID system's reliability level was disclosed, participants adjusted their reliance level more optimally than if the reliability level was not disclosed. One might assume, as Wang et al. did, that if reliability information is disclosed to participants, then belief in the reliability of the automated system should be veridical, or nearly so. Thus, one might expect that changes in reliability level should influence trust scores as well (since belief affects trust). Although Wang et al. did not report these data, they were reported in Lu Wang's Master's thesis (Wang, 2007). There was an increase in trust with reliability level in the informed condition. Trust scores were obtained using a Likert scale varying from 1 (low trust in 'unknown' feedback) to 7 (high trust in 'unknown' feedback). The mean ratings were 3.92 and 4.75 (for 67 and 80 percent reliability levels, respectively). Trust scores were also higher for those participants to whom reliability levels were disclosed (4.34) than not (2.83). Thus, it would seem that disclosure of system reliability information improves trust and reliance, and that this is due to better belief calibration. Again, these results are generally consistent with the Lee and See (2004) model.

Wang et al. (2009) also found that when system reliability levels were not disclosed, belief levels did not shift significantly, suggesting that participants could not properly gauge what the system reliability level was. However, trust scores did improve with increased system reliability in the uninformed condition (from 2.25 to 3.41; Wang, 2007). Thus, it would appear that calibrated belief in the reliability of the 'unknown' feedback was not necessary to produce changes in trust scores, and reliance. Note that this is consistent with the overall pattern of correlations among the three variables noted above. This result appears inconsistent with the Lee and See (2004) model as formulated (since there is no direct connection between automation performance feedback and trust), but the results are intriguing in that they suggest that trust can be affected by changes in system reliability without an explicit recalibration of belief.

In a study using the same simulated environment, Neyedli, Hollands, and Jamieson (2009) compared different graphical methods for representing system reliability and identity (i.e., 'friend' or 'unknown' feedback) information. Neyedli et al. manipulated display integration by combining reliability and identity information into a single display or separating the information sources. The results showed more appropriate shifts in reliance with system reliability level for integrated than for separated displays, and greater trust levels for the integrated displays. System reliability information was always disclosed in the above experiments; thus, belief estimates were not obtained. Nonetheless, these results also show the linkage between trust and reliance, consistent with the model proposed by Lee and See (2004).

In summary, Lee and See (2004) characterized a sequential relationship between belief, trust, and reliance, with trust playing an important intermediary role between explicit belief in the reliability of an automated system and reliance upon that system. The results obtained by Wang (2007; Wang et al., 2009) and Neyedli et al. (2009) make clear that disclosing system reliability affects belief and reliance. Any model of human performance in CID will need to account for these phenomena, and should clarify the mechanism by which belief, trust, and reliance are related.

A final caveat: When system reliability information is disclosed it is necessarily treated as veridical (i.e., belief equals the disclosed reliability level). One can easily imagine cases where this might not be true, so for example the operator may not trust the reliability value disclosed. Perhaps the disclosed value is not accurate.

The soldier might determine this through experience over time based on feedback from the performance of the automated system. Humans are also subject to many biases, both perceptual and cognitive, that could affect how reliability levels are interpreted. We shall develop these ideas further when we introduce our model.

## Human Performance in CID

The evaluation of human performance in combat identification seems intuitive. We want the soldier to prosecute hostile targets, and hold fire in the presence of friendly or neutral forces. However, assessing human performance in the context of an automated system complicates matters. Typically, most automation is considered either to replace the human operator, or to change the role of the human operator. Although the former is usually the intent of the engineer designing the automated system, the latter is more typically observed in practice. Thus, the human provides a governance or oversight role as the automation performs some simple, repetitive or dangerous task. In this context, one might consider the quality of human performance to reflect the ability to control an automated process, and detect and diagnose system problems, both with respect to the task being automated and the automation itself. In process control, the operator typically provides such oversight, with different task demands, skills, and abilities associated with control and diagnosis components (Wickens and Hollands, 2000). Given the complexity of process control domains, and the automated systems involved, there are many possible ways the human–automation system can fail, and thus many ways to evaluate human performance in this context.

The four-stage taxonomy of automation proposed by Parasuraman, Sheridan, and Wickens (2000) helps to shed some light on appropriate performance measures for automation in CID. They noted that automation can be applied to four function classes: information acquisition, information analysis, decision selection, and action implementation. These roughly correspond to four human information-processing stages: sensory processing, working memory, decision making, and response selection. Thus, one way of thinking about automation is that it replaces an information-processing stage performed by a human operator in a manual situation. In CID, the automated system is providing a combination of acquisition and information analysis, which Parasuraman et al. (2000) refer to as information automation. That is, the CID system senses and registers input data, and offers a classification of those data into discrete identified categories. This type of diagnostic aiding filters or focuses attention on information deemed to be of interest, and forms an inference about the state of the world. The resulting identification augments the soldier's perception and cognition. Hence, we argue that as currently implemented, CID automation meets the level of Stage 2 automation in the Parasuraman et al. taxonomy.

In CID the automated system augments, rather than replaces, an existing manual system. The aid provides another information source that can be consulted by the human operator, but the human provides the final decision. Importantly, the outcome of the human decision represents the outcome of total system performance (human plus automation). In such cases one can compare human performance alone to human performance with the decision aid. If one sees performance improve with the aid, then it is seen as useful.

Most automated decision aids operate in conditions of uncertainty and are therefore only correct a proportion of the time. This state of affairs is referred to as imperfect automation (e.g., Wickens and Dixon, 2007; Wickens, 2000). Presumably, the more reliable the automated system, the better the human–automation performance should be. In an informal metanalysis of 20 studies, human–automation performance was generally shown to improve with automation reliability (Wickens and Dixon, 2007). It is in the operator's interest to rely more on the system as its reliability increases. Wickens and Dixon also found that an automation reliability of 70 percent was the 'crossover point' below which unreliable automation was worse than no automation. Thus, importantly, people disuse automated systems if they are not sufficiently reliable.

Furthermore, it has been shown that across a wide variety of domains, reliance on imperfect automation improves overall system performance when automation reliability information is disclosed to a user (Cohen et al., 1998; Lee and See, 2004; Sheridan and Parasuraman, 2006; St. John, Smallman, Manes, Feher, and Morrison, 2005; Wang, et al., 2009). Thus, it appears that when the reliability of the automated system is disclosed to the operator, it affects how much reliance is placed upon the automation.

Automated CID technology provides an information source that is highly reliable when an RFID tag is present. However, as noted above, 'unknown' feedback as an indication of a hostile target is much less reliable. Thus, it now becomes more difficult to characterize human performance in this context. Should the soldier believe that the 'unknown' feedback indicates a hostile target? Should the soldier believe that 'unknown' feedback indicates that the likelihood of a hostile target has increased? Should the soldier ignore the 'unknown' feedback altogether since it is unreliable? It becomes less clear which of these options represents better or worse performance. The soldier must somehow evaluate the validity of the 'unknown' feedback over time to decide how much to rely on it. This evaluation likely involves a trust calibration process, and also the functional specification of trust as Lee and See (2004) have suggested. For decision aiding, we suggest here that trust in the aid is based on a history of operational experience with the aid: the decisions made and the outcomes. That is, trust is based on a reliance history. We develop these ideas further in subsequent sections.

Dzindolet, Beck, Pierce, and Dawe (2001a; see also Dzindolet, Pierce, and Beck, 2009), proposed a framework for automation use in which social, cognitive, and motivational processes all affect automation reliance. As such, the framework is similar in some respects to the Lee and See (2004) model, but differs in a number of ways. The framework emphasizes automation use and reliance, but does not incorporate feedback explicitly. In contrast, Lee and See's model represents the dynamic processes that govern trust, including feedback over time. The Dzindolet et al. framework explicitly characterizes decision aiding in the CID environment, whereas the Lee and See model is more general in its orientation. A key feature of the Dzindolet et al. framework is an explicit comparison of the perceived reliability of an automated aid and perceived reliability of manual operation. The perceived utility of an automated aid is characterized as a comparison between these two values. Perceived utility affects trust and ultimately reliance. We think that these are important ideas and develop them further below.

The Lee and See (2004) model has the advantage of considering trust as a dynamic process, and emphasizes trust calibration and appropriateness. It links performance of the entire automated system as a variable influencing trust. Thus, the level of trust is influenced by the recent performance of the automated system, by contextual variables like self-confidence, and by the recent history of trust values. Lee and Moray (1992) found evidence for this in an experiment where participants controlled a simulated process control system, and could optionally use automated aids in controlling the system. Reliance on the system and subjective trust estimates were strongly influenced by system failures. We think this perspective is valuable in developing a computational model of trust and reliance in CID.

### *CID Performance Metrics*

We can redefine *misuse* as use of the automated system when doing so leads to degraded system performance, and *disuse* as not using the automated system when doing so would lead to improved system performance. In the CID studies conducted by Dzindolet et al. (2001b; Dzindolet, Pierce, Pomranky, Peterson, and Beck, 2001c) misuse rates were computed as the error rates for trials in which the aid feedback was incorrect, or

$$Misuse = P(Error \mid \text{Aid Incorrect})$$

Disuse was defined as the error rate for the trials in which the aid feedback was correct:

$$Disuse = P(Error \mid \text{Aid Correct})$$

These metrics are somewhat informative at a gross level. For example, if misuse rates increase over time it might suggest that users are tending to increase their reliance on the system. However, there are a number of issues. First of all, the metrics do not distinguish between two error types: misses and false alarms. For example, in a disuse case, while the aid is correct in identifying a friend, the human shoots at the target (a false alarm). When the aid classifies an enemy as unknown, the human user might hold fire (a miss). Although these are qualitatively different situations, the disuse score defined above does not distinguish between them.

Second, it is difficult to compare the two values. If misuse exceeds disuse, can the soldier be said to have over-relied on the automation? This is difficult to assess, because we do not know how often the aid was correct. This method computes misuse the same way if the aid is incorrect on 50 percent or 99 percent of the trials, but the impact on performance will be very different (Wang, Jamieson and Hollands, 2008). A third related problem is that the reliability of the automated system might differ depending upon whether the aid provides 'friendly' or 'unknown' feedback. Finally, there is no indication of what optimal performance would be using these measures. Should the soldier rely more or less on the aid? In part, this would be determined by the reliability of the automated system. And as we know, the reliability of feedback from CID systems is affected by the identity feedback received. There appears to be an underlying theme here: the misuse/disuse method does not take the reliability of the automated aid into account.

It is of course possible to modify the above definitions to allow for different types of feedback, and to adjust the metrics so that they are sensitive to reliability level, and this would help to solve some of the above problems. However, even after doing so, optimal performance is still not properly defined. We prefer a different approach that allows for the definition of optimal reliance (Wang et al., 2008), and describe that approach next.

In signal detection theory (SDT) (Green and Swets, 1966; Macmillan and Creelman, 1991; Wickens and Hollands, 2000), an observer's performance is characterized by two indicators – sensitivity and response bias. SDT is typically applied to situations where an observer is presented with a situation containing either a signal or no signal and must judge whether the signal is present or not. The input channel is said to be noisy: that is, the process of determining whether the signal has occurred or not is not deterministic, but stochastic. The measures are derived from two scores: hit rate, $P(H)$, and false alarm rate, $P(F)$, based on the discrete responses of the observer.

To adapt these measures to CID, we consider two judgment types or paradigms. The first, used by Dzindolet et al., 2001b; Dzindolet et al., 2001c), and Rice, Clayton, and McCarley (2009), defines target present and target absent situations, and observers are asked to detect when a target is present. Following Macmillan and Creelman (1991), we refer to this as a *detection* task (Figure 9.1, top). In this case, the probability of a hit, $P(H) = P(\text{'Present'} \mid \text{Present})$, and $P(F) = P(\text{'Present'} \mid \text{Absent})$. For this task, whenever the target is present, it is an enemy – that is, there are no situations where a friendly target is shown. For this reason, the detection task is not very realistic. The second type of judgment is a *recognition* (or discrimination) task, and was used by Karsh, Walrath, Swoboda, and Pillalamarri (1995), Kogler (2003), Neyedli et al. (2009), and Wang et al. (2009). In this case, the observer is always presented with a target. The target is hostile or friendly, and the observer must shoot or hold fire appropriately. In this case, the probability of a hit $P(H) = P(\text{Shoot} \mid \text{Hostile})$, and the probability of a false alarm $P(F) = P(\text{Shoot} \mid \text{Friend})$, as shown at the bottom of Figure 9.1. Once the appropriate classification matrix is developed, it is then a simple matter to compute the conventional SDT measures of sensitivity and response bias, and in fact, the measures are computed identically for both detection and recognition tasks (Macmillan and Creelman, 1991, p. 5).

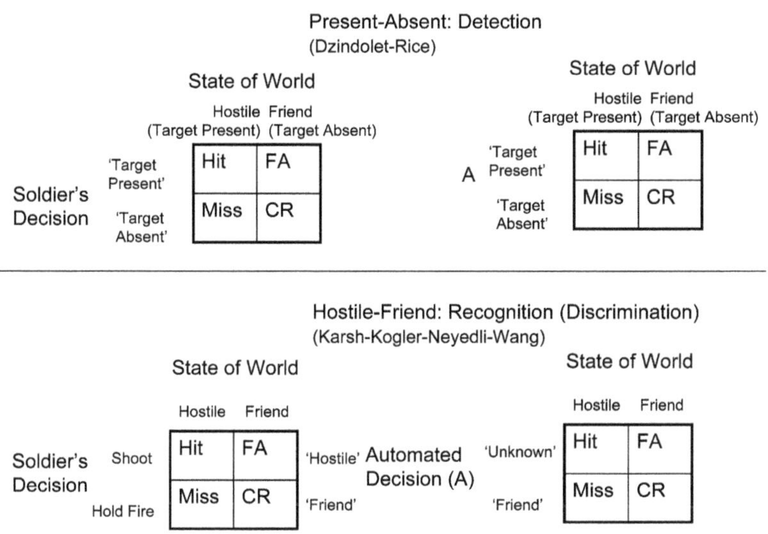


**Figure 9.1 Adaptation of SDT measures for detection and recognition CID tasks. FA = false alarm. CR = correct rejection**

Note that we can treat the behavior of an automated system itself in the same manner (e.g., Sorkin and Woods, 1985). For the detection paradigm, the automation responses ('target present' and 'target absent'), the classification matrix (Figure 9.1, top right), and $P(H)$ and $P(F)$ are all the same as for the human observer. However, for the recognition paradigm to be realistic, the automated system cannot definitively return a 'hostile' response. Thus, the automated system responds 'friend' when the RFID tag is recognized, or 'unknown' when it is not. This situation is represented on the bottom right of Figure 9.1. Thus, we define $P(H) = P(\text{'}Unknown\text{'} \mid Hostile)$ and $P(F) = P(\text{'}Unknown\text{'} \mid Friend)$. Note that the target is not positively identified by the automated system. Instead, the target is classified as 'unknown', a class that also includes allies without transponders, or who have transponders that are not detected, as noted above.

We need to consider further the meaning of target. In many automation situations, a *target* is usually a system fault, a problem the operator has to solve. Thus, the hit state occurs when there is a system fault, and the automated system says there is a fault. This is a well-defined state. Similarly, it is common in the human factors of automation literature to call the situation when the automated system says a target state is present – but it is not – as an *automation false alarm*, a situation that tends to produce the well-known *cry-wolf effect* (Bliss, 2003; Rice et al., 2009; Wickens and Dixon, 2007). See Figure 9.2 (left side).

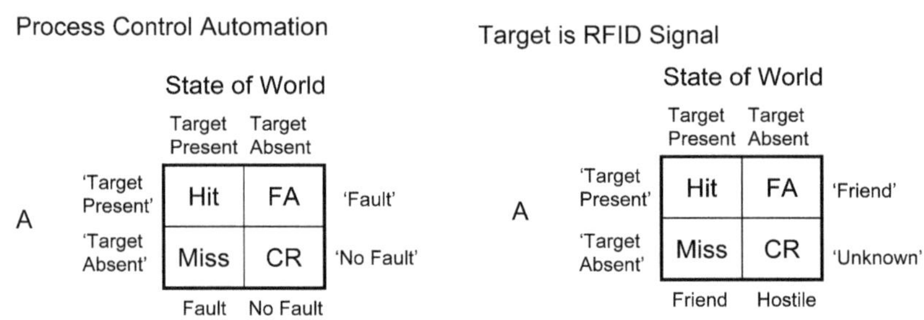


**Figure 9.2 Targets are fault states in process control automation (left). In CID if the target is considered the RFID signal, then the target is a 'friend' classification (right). A = automated system. FA = false alarm. CR = correct rejection**

What is the target in the CID context? We could say that for the automated CID system, the target is the RFID signal from the transponder, as shown on the right of Figure 9.2. While this has the advantage of clear definition, the aid identifies a friendly agent, not the target the soldier must prosecute. For the soldier, as discussed above, and depicted in Figure 9.1 (left side), the target is typically considered to be an enemy. Thus the RFID 'target' for the automated aid corresponds to the friend classification – just the opposite of the target in the human sense! On the other hand, if we consider the target for the automated CID system to be the enemy, it is problematic for other reasons. A hit occurs when the enemy soldier is present and the aid says 'unknown.' The automated system cannot positively identify this state. Instead, its classification of 'unknown' represents a range of possibilities.

There appears to be no good solution to the target-definition problem. It is essentially a problem of diagnosticity – what the automated system can provide in terms of output classes does not correspond perfectly to the classes needed to perform the task. We will therefore appeal to consistency with the soldier's task and define the target of an automated CID aid as an enemy soldier, and therefore $P(H) = P(\text{'Unknown'} \mid Hostile)$ (Figure 9.3). We will see later that most information sources available to the soldier performing combat identification are similarly imperfect in terms of diagnosing the correct identity.

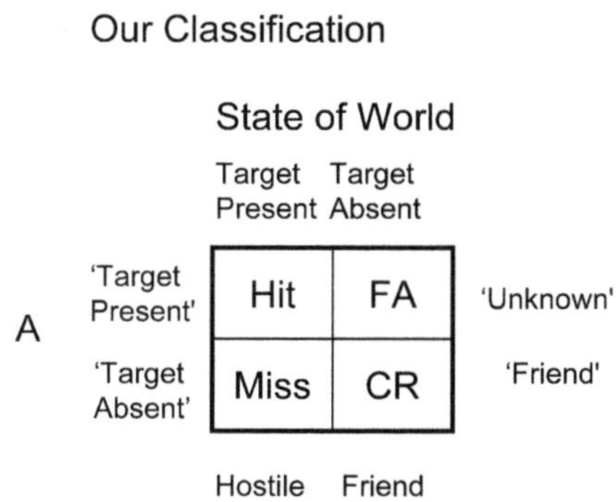


**Figure 9.3 Definition of target for CID model. A = automated system. FA = false alarm. CR = correct rejection**

SDT models have been used to characterize human performance with automation. For example, Sorkin and Woods (1985) treat the automated system and human operator each as sampling from a noisy input vector *X*. Importantly, Sorkin and Woods assume that the operator will generally observe the input channel only when alerted by the automated monitor stage. This is typical of most automation in industrial process control applications. However, it does not properly capture the CID situation, in which the user samples the available visual cues along with the output of the automated CID aid. The Sorkin and Woods approach also does not capture the fact that the reliability of CID system output depends on the information that it provides. If the system is highly reliable with friend feedback, it is not really sampling from a noisy channel in this case (i.e., noise is virtually zero). However, when one considers the unreliable 'unknown' feedback, the Sorkin and Woods model could represent a valid approach for CID if suitably adapted. We argue that for CID, it is important to view the human operator as sampling from a set of input channels from the environment, including the CID system. In a sense, the model we will develop represents an adaptation of the Sorkin and Woods approach to CID, but importantly also considers the user's trust in the automation, and how that trust interacts with human reliance and sensitivity.

When a CID system provides feedback on target identity, it is likely to affect how the user responds. Thus, we are likely to obtain different *P(H)* and *P(F)* scores from the human user given 'unknown' or 'friend' feedback from the aid. When users receive 'unknown' feedback,

$$P(H) = P(Shoot \mid \text{`Unknown'} \cap Hostile)$$

and

$$P(F) = P(Shoot \mid \text{`Unknown'} \cap Friend)$$

*P(H)* and *P(F)* can be similarly computed for the 'friend' feedback situation, by replacing 'unknown' with 'friend', thus,

$$P(H) = P(Shoot \mid \text{'Friend'} \cap Hostile)$$

and

$$P(F) = P(Shoot \mid \text{'Friend'} \cap Friend)$$

However, since 'friend' feedback is extremely unlikely with a hostile target, a hit is unlikely. Thus, with 'friend' feedback, *P(H)* is likely to be near zero. It is therefore difficult to derive measures of sensitivity or response bias with 'friend' feedback in practice.

Using conventional SDT measures, the response bias parameter $\beta$ can be obtained from *P(H)* and *P(F)* scores. In the 'unknown' feedback context, we will refer to this as $\beta_{unknown}$. For example, if 'unknown' feedback indicates that a target is likely to be hostile, one would expect that a user would likely show a reduced $\beta_{unknown}$ value and become more liberal, relative to an unaided situation. The more reliably the 'unknown' feedback indicates a hostile target, the more likely it is that $\beta_{unknown}$ should decrease.

The SDT approach allows for a computation of an optimal response bias value based on probabilities (Green and Swets, 1966; Macmillan and Creelman, 1991). Thus, in our terms,

$$\beta_{optimal} = P(Friend) / P(Hostile)$$

The likelihood of friendly and hostile targets is clearly of some importance to human performance in combat and is likely to vary across levels of conflict and moment by moment in the battlefield. An SDT approach can also incorporate the influence of payoffs, or the values and costs of the four possible outcomes: hits, misses, false alarms, and correct rejections. Thus,

$$\beta_{optimal} = P(Friend) / P(Hostile) * ([V(C) + C(F)] / [V(H) + C(M)])$$

The effect of varying the probability of friendly and hostile targets is an interesting question its own right, but most studies to date have examined the effect of varying the reliability of an automated system upon human reliance using that system (e.g., Dzindolet et al., 2002; Neyedli et al., 2009, Rice et al., 2009).We are particularly interested in the situation in which the user receives 'unknown' feedback from the automated aid. In this case, we can redefine optimal response bias as:

$$\beta_{optimal} = P(Friend \mid \text{'Unknown'}) / P(Hostile \mid \text{'Unknown'}) * ([V(C) + C(F)] / [V(H) + C(M)])$$

Note that this value will decrease as the 'unknown' feedback from the CID system becomes more reliable, since the numerator will decrease and the denominator will increase. Payoffs, as more enduring costs and values of outcomes, are less likely to vary with 'unknown' versus 'friendly' feedback.

The appropriateness of reliance is defined as the difference between a user's response bias and this optimal value. When the empirical $\beta$ value is greater than optimal, it is evidence for misuse; conversely, when it is less than the optimal value, it is evidence of disuse. One can clearly see the advantage of this approach versus extant misuse and disuse approaches.

While response bias is important for assessing reliance, SDT also provides a sensitivity measure that can be used to assess CID performance. Sensitivity measures the resolution of the detector by assessing the ability of the detector to discriminate between the signal (in our case a hostile target) and noise (i.e., friendly soldiers, civilians) (Green and Swets, 1966; Macmillan and Creelman, 1991). More formally, sensitivity is the difference in standard deviation units between the noise distribution and the signal plus noise distribution along the continuum of neural activation *X*. A strong signal will produce more activation, making the signal more easily discriminated from the lower level of activation when only noise is present. In practice we can calculate a sensitivity parameter as:

$$d' = z(F) - z(H)$$

where $z(F)$ and $z(H)$ are the $z$ scores that correspond to these right-tail $p$-values represented by $P(F)$ and $P(H)$. This measure is useful as it is affected by physical or display properties like the salience of the signal or properties of the observer (e.g., through training) (Wickens and Hollands, 2000). For our purposes, it can be used to check whether manipulations designed to affect reliance have an effect (beneficial or not) on the salience of the target or the resolution of the detector.

A similar approach was espoused by Sheridan and Parasuraman (2000), who used expected-value decision theory to help determine whether a human or automated system is best for a failure-detection task. Their value $K$ is equivalent to $\beta_{optimal}$, but does not require the same parametric assumptions about the distributions of $X$. Thus, if suitably adapted to CID, the Sheridan and Parasuraman approach offers a non-parametric equivalent to our approach.

To illustrate the utility of SDT measures, we consider here data collected by Karsh et al., (1995). Participants were shown slides of target vehicles, which were either friendly (US Army) or enemy tanks. After training, participants had to classify the vehicles as friendly or enemy, either with or without assistance from an automated system. Errors were classified as fratricide or suicide (not shooting when target was hostile).

From these data, we were able to compute sensitivity and reliance scores. Figure 9.4 shows the data classified in the matrix from Figure 9.3. Note first that given the high reliability of the automated system, its sensitivity was very high, and its $\beta$ value was set low. In contrast, human sensitivity without the automated system was much poorer ($d'$ = 3.03 vs. 1.49). Without the automated system, $\beta_{optimal}$ is determined by probabilities (there were no explicit payoffs in the Karsh et al. (1995) experiment) and would therefore be set to 1, given that hostile and friendly targets were equally probable. When we consider human performance with the automated system, we can now compare our obtained reliance on the automated system with $\beta_{optimal}$. We calculated $\beta_{optimal}$ as .104 (very liberal) (see lower right of Figure 9.4). This is because in the Karsh et al. experiment the probability of a hostile target given 'unknown' feedback was much greater than the probability of a friendly given 'unknown' feedback. To compute the empirical beta value (representing obtained reliance), we used *P(H)* and *P(F)* to construct the classification matrix. More correctly, *P(H* | 'Unknown') and *P(F* | 'Unknown') should be used; however, Karsh et al. (1995) did not report these values. Nonetheless, *P(H* | 'Friendly') and *P(F* | 'Friendly') were likely very small values (since it is unlikely that participants would classify a target as enemy given highly reliable 'friendly' feedback), and so the values in the classification matrix would likely not be that different. Participants in the assisted condition (using the automated aid) showed an average $\beta$ value of 1.27, less than in the manual condition, but still well above the low optimal value, an example of the sluggish beta phenomenon (Wickens and Hollands, 2000). The implication is that participants were unwilling to lower their criterion sufficiently and declare a target hostile despite the highly reliable feedback from the automated system. Sensitivity improved as well, with a $d'$ of 2.36, about halfway between manual performance

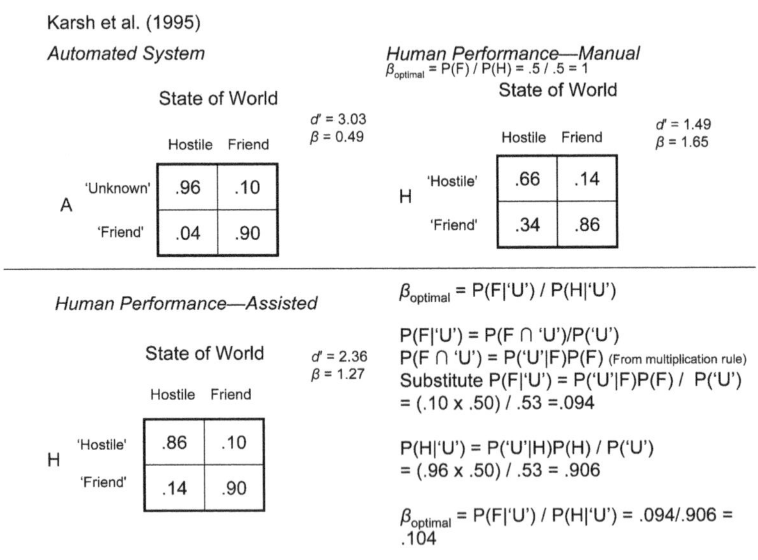


**Figure 9.4 SDT analysis on data from Karsh et al. (1995). For calculation on bottom right F = friend and H = hostile**

and the sensitivity of the automated system. In summary, it is clear that providing a highly reliable automated CID system was useful in this case, with the combined sensitivity of the human-automated system better than that of human observers without such a system. However, performance was less than that of the automated system, and reliance was much more conservative than it should have been.

In another study, Kogler (2003) had participants perform CID with an imperfect aid. The participants wore goggles that limited the level of transmissivity to 75, 10 or 2 percent. These levels corresponded to vision with clear glasses, ballistic-protection sunglasses, and vision at dusk respectively). The participants either used no automation or automation that was 60 percent or 100 percent reliable.

Kogler (2003) showed that reduced transmissivity increased friendly-fire engagements while increased reliability reduced friendly-fire incidents. We performed an SDT analysis (Figure 9.5) on the available data, which showed that increasing reliability, $F(2,22) = 48.66, p < 0.01$, and transmissivity, $F(2,22) = 4.86$, $p < 0.05$, increased sensitivity. Reliability level also affected response bias, $F(2,22) = 4.74$, $p < 0.05$; participants had a more liberal $\beta$ with no aid and a 60 percent reliable aid. There was no main effect of transmissivity on response bias. Thus, using an SDT analysis, even with limited data, we dissociated the effect of reduced visibility for sensitivity versus response bias. We calculated $\beta_{optimal}$ using the same method we used for analyzing the data from Karsh et al. (1995), separately for manual control ($\beta_{optimal} = 0.49$), 60 percent reliable aid ($\beta_{optimal} = 0.20$), and 100 percent reliable aid ($\beta_{optimal} = 0.00$) conditions. We once again used $P(H)$ and $P(F)$ to construct the classification matrix and compute the obtained reliance since $P(H$ | 'Unknown') and $P(F$ | 'Unknown') were not reported. Given that the average $\beta$ values were 1.50, 1.18 and 2.17 in the manual control, 60 percent and 100 percent reliable conditions respectively, the participants were well above the optimal value in all conditions.

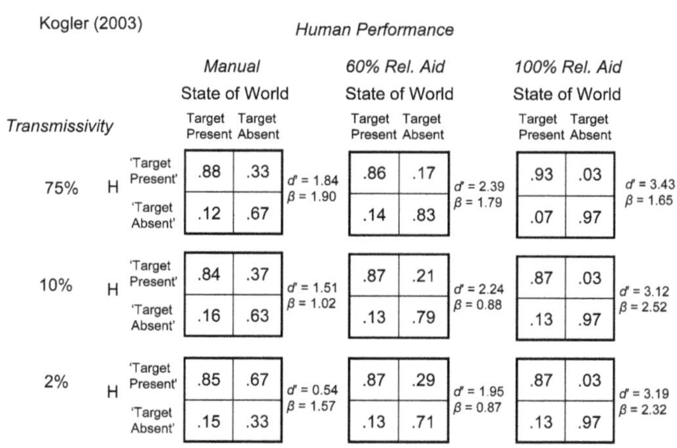


**Figure 9.5 SDT analysis on data from Kogler (2003)**

## Sampling Model for Combat Identification

We take as a starting point that reliance on an automated system is essentially a decision about risk. Thus, from classical expected utility (von Neumann and Morgenstern, 1944), we characterize the utility of a decision as a ratio between the values of two alternatives. In objective terms, utility $U$ is therefore $R_a / R_m$, where $R_a$ and $R_m$ are the reliabilities of the automated and manual systems, respectively. However, we can consider that the subjective estimates of reliability are not veridical, and therefore perceived utility, or belief $B$, is equal to:

$$B = \Psi(R_a) / \Psi(R_m) \quad (1)$$

where $\Psi(R_a)$ and $\Psi(R_m)$ indicates the subjective estimates of automated and manual systems, respectively. Following extant work on subjective utility (Bernasconi, Choirat, and Seri, 2008; Prelec, 1998; Tversky and Kahneman, 1992) we treat the relation of the two subjective reliabilities as a ratio (rather than a difference, as Dzindolet et al., 2001a proposed).

System reliabilities represent probabilities. They are computed by taking the total number of system successes over a total number of events. As such they represent a *proportion* of total events. When human observers judge proportions, they make systematic biases. The form of the bias for proportion judgments has been well characterized (Hollands and Dyre, 2000; Karmakar, 1978; Spence, 1990). Specifically,

$$P(R_.) = R_.^b / [(R_.^b) + (1 - R_.)^b] \quad (2)$$

When plotted against $U$, this function will produce patterns showing overestimation of proportions less than half and underestimation for proportions greater than half (when $b < 1$). The value of $b$ varies with the nature of the perceptual stimuli, but for spatial areas or numerals (the typical methods for displaying system reliability information) $b$ is typically less than unity (Gescheider, 1988; Hollands and Dyre, 2000).

Social and cognitive biases such as bias toward automation, self-serving biases (Dzindolet et al., 2002), and bias resulting from comments from other team members (Lee and See, 2004) can all influence subjective estimates of reliability. We treat all such biases $e$ as simply inflating or deflating the subjective estimate. Thus we assign $\Psi(R_.) = P(R_.) + e_.$. To this point we have considered the reliability of a system, be it manual or automated. We now propose that such systems are comprised of at least one perceptual channel that can be sampled. Each individual channel also has a reliability, defined as above, except that we also include a new parameter $i$, referring to the identity of the channel's output ('hostile' or 'friendly'). This is because in many cases the reliability of a system's output for one identity is different from that for the other.

A channel's perceived reliability will likely also be influenced by salience. Bryant and Smith (2009) conducted an experiment examining decision making in CID. In a simulation, the participant was shown a human figure moving through urban terrain. The subject's task was to engage (i.e., shoot) only those figures that were enemies. Friendly and enemy forces were distinguishable by differences in uniforms or equipment, but these characteristics were uncertain (i.e., not 100 percent reliable). Two factors were varied: whether the uncertain characteristic was visual or behavioral, and whether it was salient or not. The results indicated that when a salient visual or behavioral characteristic of a friend was uncertain, participants were more likely to false alarm (engage the friendly target) than when a non-salient characteristic was uncertain. The implications for our model are that more salient channels are likely to be perceived as more reliable.

We assume that the various information channels vary in diagnosticity. It is possible that a sample returns non-diagnostic evidence (e.g., a green uniform could provide evidence consistent with 'friend' or 'hostile'). An important instance of lack of diagnosticity for our purposes is when a CID system returns the classification 'unknown.' The strength of the diagnosticity $D$ can be used to weight input channels differently (e.g., if the correlation is known to be weak, the impact of the channel can be reduced). One can obtain the information value by taking the product of diagnosticity and reliability (Wickens and Hollands, 2000). This can be done by replacing $R$ in the above equations with $I$ where $I = D \times R$.

Thus we can re-express Equation 1 as

$$B = \Psi(I_a) / \Psi(I_m) \tag{3}$$

We propose here a model of combat identification in which the human observer samples from the channels available to determine a target's identity. Thus, the observer judges the identity of a target by examining the CID display (display elements showing identity information), the target, and any other relevant information sources. Multiple information channels, $C_n$, compete for the observer's attention. Examples include the color, texture, or pattern of a uniform or other clothing, the shape, color, or pattern of a helmet, weapon characteristics, or information provided by the CID system. Each time a channel is sampled, it returns two values. One is the identity value $i$, which is either 1 or -1. The likelihood of a positive or negative identity value can be adjusted based on the likelihood ratio $P$(Hostile) / $P$(Friendly). The second is evidence value $E$, the product of a random value sampled from a uniform distribution between 0 and 1. Note that if a channel does not return an identity, $E = 0$. All available channels are assumed to be sampled, but the weighting $\Psi(I)$ is applied to each channel to reflect any biases toward or away from a channel. Note that different $\Psi(I)$ values are applied depending on the value of $i$. Thus, the expression

$$O_c = E_{c,t} \times i \times \Psi(I)_i \tag{4}$$

represents the output $O$ as a function of evidence $E$, identity $i$, and the weighted belief function. $O$ is a function of the particular channel $c$ and a finite number of samples, $t$. Given that $i$ can be positive or negative, $O_c$ will be signed similarly and can vary between -1 and +1. The number of samples is a free parameter, whose value can be estimated by fitting the model to empirical response time data.

The sum of the outputs from all channels is computed as $X_t = \Sigma(O_c)$ (Figure 9.6). A threshold criterion value is set as $X_c$. If $X_t$ is greater than $X_c$, then the action is to shoot the target. If $t$ samples are reached but $X_t < X_c$, the action is to hold fire.

It should be obvious that our summed output value $X$ (which has both random and systematic elements) and its comparison to a criterion value $X_c$ can be used to generate data amenable to SDT analysis. The model will produce the discrete responses 'shoot' or 'hold fire' in response to discrete states of the world ('friendly' or 'hostile'). The classification matrix shown in Figure 9.3 can be used to classify the data and compute the relevant probabilities. Sensitivity and response bias parameters can be calculated in the manner described above. Other parameter values can be estimated by fitting the model to experimental data. We plan to undertake these efforts in future.

Note that for a more reliable channel, its output $O$ will tend to be more extreme (closer to -1 or 1). This is because the value of $O$ will tend to increase with $\Psi(R)$, which increases with the channel's reliability (since $P(R)$ is monotonic and $e$ only affects the intercept, not the slope). The channel output will also increase as the diagnosticity of the channel increases as well as when the number of samples, $t$, increases.

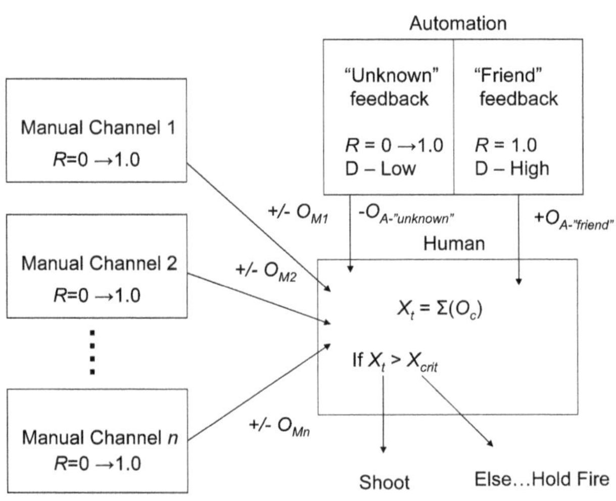


**Figure 9.6 Sampling model of CID. Manual channels could be adapted to reflect different feedback types as shown above for automated feedback**

In most cases, the perceived reliability $\Psi(R)$ of a channel will be learned through training or experience and therefore stored in the soldier's long-term memory. However, in some situations (e.g., with an automated CID system) channel reliability information can be updated over time, and therefore maintained in working memory. A channel need not be associated with a single location in the visual array, or even a single modality; for example, identity information might be found in one spatial location and associated reliability information might be obtained in another. Alternatively, identity information might be obtained from a visual source and information about reliability heard auditorially. Thus, for example, a soldier might be cautioned by a team member not to assume that an article of clothing or vehicle logo implies a hostile target. It is also possible that soldiers are misinformed about the true reliability of an automated system. In this case, the process of learning requires a recalibration of belief from performance feedback. We consider this process further in the next section.

Sets of channels can be constructed for analytic purposes by taking averaged $\Psi(R)$ functions. Thus, for example, $\Psi(R_m) = \Sigma\, \Psi(R) / n$. These sets can be averaged to represent the perceived reliability of automation or perceived self-confidence (belief in one's ability to sample accurately relevant cues in the environment). Consider Equation 1: When $B > 1$, then the bias is toward weighting the information sampled from the automated system more highly than from other sources. Note that a similar ratio can be computed for any pair of channel sets. Thus, for example, we could compute belief in the context of a set of visual cues versus verbal input received from other team members. Note that Equations 1 and 3 do not distinguish between reliabilities associated with friend and unknown feedback. However, here we are not concerned with the identity information as much as the reliability of an entire system, regardless of the reliability of the specific identity information it provides.

## Trust

Lee and Moray (1992) and Lee and See (2004) view trust as a function of system performance. Trust is adjusted based on recent performance experiences: it evolves based on feedback from performance success or failure. In CID, better performance occurs when the soldier is better able to distinguish between friend and foe (sensitivity) and relies on the automated system most appropriately. This is why it is important to define an optimal measure of reliance. Trust is different from belief. As discussed previously, belief is about the perceived reliability of information sources; trust reflects how much reliance the user puts on the automated system to optimize total system performance. Trust evolves over time and must be continuously recalibrated. Although belief can change over time, it is likely to change more slowly.

Following Lee and Moray (1992), we argue that trust is best defined as a transfer function, using an autoregressive moving average (ARMAV) form. The

Lee and Moray expression represents a first-order lagged system. In the simplest case, the function

$$T_t = \Phi_1 T_{t-1} + P_t + \Phi_2 P_{t-1} + a_t \tag{5}$$

could be used, where $T$ is trust, $P$ is performance, $t$ is a discrete time interval, $\Phi_1$ and $\Phi_2$ are time constants, and $a$ is random noise. Lee and Moray also incorporated system faults into the forcing function, requiring an additional time constant and weighting parameters to capture the relative contributions of performance and fault components. In CID, responses to incorrect feedback could be modeled similarly to system faults in Lee and Moray's model. Lee and Moray found that trust dropped quickly after a system fault, and stayed low, whereas performance levels reverted to pre-fault levels more quickly. Trust could therefore be re-expressed as:

$$T_t = \Phi_1 T_{t-1} + A_1 P_t + A_1 \Phi_2 P_{t-1} + A_2 F_t + A_2 \Phi_3 F_{t-1} + a_t \tag{6}$$

Where $\Phi_3$ is the additional time constant and $A_1$ and $A_2$ are the weighting parameters, and $F$ is incorrect feedback (automation false alarm or miss).

However, applying the model as it stands to CID represents a set of challenges. $P$ and $F$ are considered to be continuous variables (Lee and Moray modeled continuous performance data) and at the event level our data are binary and discrete (shoot or hold fire, 'friendly' or 'unknown' feedback). One solution is to use longer time intervals, effectively characterizing performance over a set of discrete events. This would enable us to define $P$ as a function of sensitivity and deviation of reliance from optimality. These could be combined or treated as separate sensitivity and bias parameters. We aim to pursue these modeling efforts in future work.

Note that we keep trust and performance generic in that channels are sampled and reliance is placed upon them in terms of their reliabilities and biases towards or away from the channel. There is no reason why our approach could not be used to characterize trust and reliance on any single channel or set of channels. So for example, one could contrast CID performance for a soldier working solo or as a member of a team, with the various team members serving as information channels. Reliance on team could be contrasted with self-reliance. Other scenarios are easily constructed (e.g., working with NGOs versus members of one's own unit, working with an automated system operated by another team member versus oneself; working with an IFF system versus SA system, etc.).

## Predictions

The model makes a number of predictions with respect to human performance (sensitivity and response bias). We describe several here, but the list is not exhaustive. We encourage the reader to develop others.

The first prediction is that reducing the time to sample a channel should decrease sensitivity. There is evidence consistent with this prediction. Neyedli (2009, Experiment 2) found that sensitivity was lower with limited time to sample a combat scene. In her experiment, which required participants to decide if a target was friendly or hostile, when participants viewed a combat scene for 400 ms they had lower sensitivity that when viewing a combat scene for 800 ms. The sensitivity scores from this experiment were much lower than those obtained by Neyedli et al. (2009), whose participants took approximately 3 seconds to respond. A similar prediction is reducing the salience of the signal for a given sample should decrease sensitivity. As demonstrated with the data from Kogler (2003), decreasing lens transmissivity reduced human sensitivity in CID.

The second prediction is that using display formats that improve detection of identity or reliability information should improve sensitivity. However, if more samples are required to obtain an estimate of system reliability, so in a time-pressured situation, sensitivity should be reduced as the participant would have fewer samples remaining to sample the target. For example, a number of recent studies have suggested that a random mesh indicator (Figure 9.7, left) provides an effective 'at a glance' representation of proportion (e.g., Feldman-Stewart, Brundage, and Zotov, 2007; Finger and Bisantz, 2002), more effective that conventional graphical formats. The proportion of the mesh that is filled in (either with shading or color) represents the proportion or probability in question. Neyedli et al. (2009) compared this mesh format to a pie chart representation (Figure 9.7), to represent the reliability of a CID display. System reliability varied trial by trial in their experimental simulation. They found that the mesh display produced greater sensitivity than the pie display.

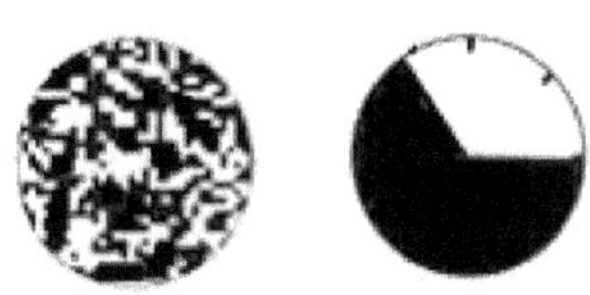

**Figure 9.7 A random mesh indicator of proportion (left) and a more conventional pie chart indicator of proportion (right)**

The third prediction is that disclosing automation reliability levels to users should lead to more optimal reliance. This is because the operator will be better able to weight evidence from the automated channel. In turn, more optimal reliance should increase performance of the human–automation team. Less effective formats for displaying system reliability should also lead to less optimal response bias because there is more variability (greater uncertainty) about the system reliability level. Neyedli et al. (2009) also manipulated whether system reliability and identity information were integrated into the same display or placed in two separate locations. In the latter case, this introduces separate sampling channels for reliability and identity information. The model would predict that if on a given sample reliability information was not obtained, then $E = 0$ and the channel returns no information. Thus, reliance will tend to be reduced to chance levels. Neyedli et al. found that reliance levels were more optimal with the integrated than the separated display format.

The fourth prediction is that varying reliability levels should affect reliance levels. Wang et al. (2009) and Neyedli et al. (2009) both found evidence in support of this prediction. Dzindolet et al. (2001c) did not, but did not analyze their data so as to separate reliance levels from sensitivity measures, instead contrasting disuse and misuse. Our reanalysis of the experimental data obtained by Kogler (2003) using SDT to separate reliance levels from sensitivity showed varying reliability level also affected reliance.

A fifth prediction relates to belief and trust. Disclosing automation reliability levels to users should affect their beliefs about the reliability of the automated system. This should lead to improved trust calibration. In turn, the improved trust calibration should help in the calibration of reliance, leading to beta settings closer to optimal. This sequential process hinges on being able to sample reliability information accurately. If reliability information is communicated effectively it may overcome some of the biases affecting perceived reliability of the system. Wang (2007) found improved trust when automation reliability was disclosed to the participants. If, in contrast, soldiers are misinformed about reliability levels, then the trust calibration process will take longer, and reliance will be sub-optimal for a longer period.

A sixth prediction also relates to the sequential relationship between belief, trust, and reliance. As performance improves, error is reduced, belief and trust increase, and the cycle perpetuates. At this point belief, trust, and reliance are all well calibrated, leading to reduced sampling time (because more reliable channels are sampled more frequently), improved sensitivity, and more optimal reliance. As soldiers detect discrepancies between the identity information provided by a channel and true identity through feedback, perceived reliability and trust are adjusted and reliance is shifted, leading to improved performance.

Performance of the automation is an important factor in trust of the automation; however, the automation must be used to gain information about the performance. Thus a seventh prediction is that if initially the user is biased against automation, they will reduce the weight of those channels associated with the automation, based on their perceived reliability $\Psi(R_a)$. This may lead to disuse of automation even when reliable as the user has not experienced the benefit of use of the automated system. On the other hand, if the operator uses the automation, they will be better able to calibrate their trust level based on prior performance on subsequent occasions. This should lead to better human–automation performance through more appropriately calibrated reliance behaviors. And indeed, Wang et al. (2009) has demonstrated that trust is correlated with reliance.

## Implications and Conclusions

Trust is required in teams to help achieve common goals. The task of CID is challenging and requires a high level of coordination among team members. We focused on the automated team member in CID, but we believe that many of the ideas could easily be extended to trust among the human team members. As automated CID systems require active involvement of the operator (as opposed to automation for supervisory control), the role of automation as a team member becomes more important for modeling. Interpersonal trust and human–automation trust is similar; however, operators are likely to have biases regarding their interactions with automation.

We used SDT to quantify reliance and performance of the human–automation team, and to clarify the measurement of human performance with automation. This approach affords separation of measures for sensitivity and automation reliance (response bias), and also provides a definition of optimal reliance, helping to clarify when automation disuse or misuse can be said to have occurred.

We introduced a sampling model to describe and predict performance and trust in the human–automation team. The operator samples information from a number of channels including information from the automated system. If the information from the samples accumulates sufficient evidence of a hostile target, the decision is made to shoot. Importantly, this evidence is weighted based on the perceived reliability of the channel. The perceived reliability is affected by such factors as biases towards the automation and, indirectly, operator self-confidence. A belief score can be calculated for two sets of channel by comparing the weighted reliabilities of each set. This comparison can be particularly useful in examining the belief between automated and manual channels.

Trust plays an intermediary role in the relationship between belief and reliance (Figure 9.8). More stable beliefs arise as the perceived reliabilities of channels or systems increase. Beliefs then influence trust that also varies as a function of system performance and failures. Trust that is calibrated to system performance can help users rely on the system optimally. More appropriate reliance should lead to better performance. Based on feedback, the operator may adjust their beliefs and trust levels regarding the system.

A number of implications arise from the model. Biases play an important role in weighting perceived reliability, and therefore belief: thus, organizational, cultural, and team factors act as a context in which beliefs reside. As such, considering an automated system as part of a team will assist in the assessment of automation performance. Second, the model highlights the role of the operator in the CID task by treating automation as only one of the possible channels from which the soldier can sample. The inclusion of other channels allows the model to

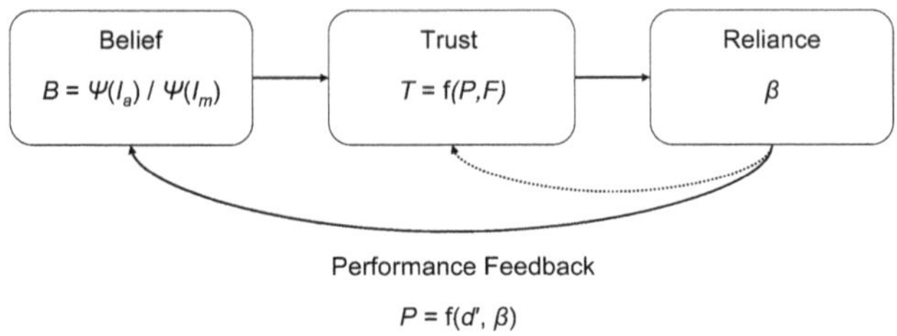


**Figure 9.8 Relationship between belief, trust, and reliance**

characterize interventions not specifically related to CID technology (e.g., those related to training, tactics, and procedures).

The use of both sampling and perceived reliability in the model acknowledges the importance of bottom-up and top-down factors affecting performance. First, a sampling process suggests that the user needs to perceive information about the target and anything that degrades the quality of or reduces the amount of sensory evidence will impair performance. Such factors are likely to affect sensitivity. Second, the inclusion of perceived reliability recognizes that calibrated knowledge of channel reliability will affect reliance behavior.

It is important to recognize that given the contingent relationship between the human and the automated system, the combined human-automated system's sensitivity and response bias are not independent. The automated system has a certain sensitivity and response bias, and this will affect the human observer's sensitivity and the reliance the human observer places upon the automated system. This stands in contrast to the use of SDT to characterize performance in psychophysical tasks.

Consider an example to illustrate this point. A human observer is presented with a scene containing either an enemy or friendly target. The observer samples the available cues in the environment. If the decision criterion is set low (liberal), this implies that the observer is more likely to score hits and make false alarms (and less likely to miss targets or make correct rejections). However, once we introduce an automated system, we now have to consider whether the soldier shoots or not in the context of certain feedback from the system. If users are informed of the aid reliability level with 'unknown' feedback, response bias decreases (becomes more liberal) as the reliability of the system increases. Thus, they are more likely to shoot. However, the likelihood of the combination of 'unknown' feedback and friendly also decreases as the system becomes more reliable. Therefore, it becomes more difficult to actually false alarm even as the user lowers the criterion.

ROC curves can be generated for the combined system, showing the interrelation between the two measures. It is therefore likely that performance will be constrained to certain regions of the ROC space. A similar approach conducted by Sorkin and Woods (1985) showed ROC curves that were constrained. However, they modeled automated systems for process control where the human observer does not sample unless alerted by the automated system. In future work we plan to use the model to explore these predictions where the operator is more actively involved, as is the case in the context of CID.

In conclusion, we believe that our modeling approach fleshes out earlier descriptive models like those proposed by Lee and See (2004) and Dzindolet, Beck, et al. (2001a). Figure 9.8 illustrates how the model provides specific computational mechanisms underlying belief, trust, and reliance. We believe that future work should assess the validity of such mechanisms for characterizing and predicting human trust and reliance on CID systems.

## References

Adams, B. and Webb, R. (2003). *Trust development in small teams.* Defence Research and Development Canada – Toronto Contract Report, CR-2003-016. Guelph, Ontario: HumanSystems Incorporated.

Ajzen, I. and Fishbein, M. (1980). *Understanding attitudes and predicting social behavior*. Englewood Cliffs, NJ: Prentice-Hall.

Baba, M. L., Falkenburg, D. R., and Hill, D. H. (1996). Technology management and American culture: Implications for business process redesign. *Research Technology Management*, *39*, 44–54.

Bainbridge, L. (1983). Ironies of automation. *Automatica*, *19*(6), 775–779.

Bernasconi, M., Choirat, C., and Seri, R. (2008). Measurement by subjective estimation: Testing for separable representations. *Journal of Mathematical Psychology*, *52*, 184–201.

Bliss, J. P. (2003). Investigation of alarm-related accidents and incidents in aviation. *International Journal of Aviation Psychology*, *13*, 249–268.

Bowers, C. A., Oser, R. L., Salas, E., and Cannon-Bowers, J. A. (1996). Team performance in automated systems. In R. Parasuraman and M. Mouloua (eds), *Automation and human performance: Theory and applications* (pp. 243–263). Mahwah, NJ: Erlbaum.

Bryant, D. J. and Smith, D. G. (2009). *Impact of uncertain cues on combat identification judgments*. Defence Research and Development Canada – Toronto Technical Report, TR 2009-027.

Cohen, M. S., Parasuraman, R., and Freeman, J. T. (1998). Trust in decision aids: A model and its training implications. In *Proceedings of the 1998 Command and Control Research and Technology Symposium.* Washington, DC: Command and Control Cooperative Research Program.

Dirks, K. T. and Ferrin, D. L. (2001). The role of trust in organizational settings. *Organizational Science*, *12*(4), 450–467.

Dzindolet, M. T., Beck, H. P., Pierce, L. G., and Dawe, L. A. (2001a). *A framework of automation use.* Army Research Laboratory Technical Report, ARL-TR-2412.

Dzindolet, M. T., Pierce, L. G., Beck, H. P., Dawe, L. A., and Anderson, B. W. (2001b). Predicting misuse and disuse of combat identification systems. *Military Psychology*, *13*, 147–164.

Dzindolet, M. T., Pierce, L., Pomranky, R., Peterson, S., and Beck, H. (2001c) Automation reliance on a combat identification system. In *Proceedings of the 45th Annual Meeting of the Human Factors and Ergonomics Society* (pp. 532–536). Santa Monica, CA: Human Factors and Ergonomics Society.

Dzindolet, M. T., Pierce, L. G., Beck, H. P., and Dawe, L. A. (2002). The perceived utility of human and automated aids in a visual detection task. *Human Factors*, *44*, 79–94.

Dzindolet, M. T., Pierce, L. G., and Beck, H. P. (2009). An examination of the social, cognitive, and motivational factors that affect automation reliance. In D. H. Andrews, R. P. Herz, and M. B. Wolf (eds), *Human Factors Issues in Combat Identification* (pp. 277–297). Farnham, UK: Ashgate.

Endsley, M. R. and Kiris, E. O. (1995). The out-of-the-loop performance problem and level of control in automation. *Human Factors*, *37*, 381–394.

Feldman-Stewart, D., Brundage, M. D., and Zotov, V. (2007). Further insight into the perception of quantitative information: Judgments of gist in treatment decisions. *Medical Decision Making*, *27*, 34–43.

Finger, R. and Bisantz, A. M. (2002). Utilizing graphical formats to convey uncertainty in a decision making task. *Theoretical Issues in Ergonomics Science*, *3*, 1–25.

Gescheider, G. A. (1988). Psychophysical scaling. *Annual Review of Psychology*, *39*, 169–200.

Gizewski, P. and Rosteck, M. (2007). Toward a JIMP-capable land force. *Canadian Army Journal*, *10*, 55–72.

Green, D. M. and Swets, J. A. (1966). *Signal detection theory and psychophysics*. Oxford, UK: John Wiley.

Hollands, J. G. and Dyre, B. P. (2000). Bias in proportion judgments: The cyclical power model. *Psychological Review*, *107*, 500–524.

Jian, J., Bisantz, A. M., and Drury, C. G. (2000). Foundations for an empirically determined scale of trust in automated systems. *International Journal of Cognitive Ergonomics*, *4*, 53–71.

Tversky, A. and Kahneman, D. (1992). Advances in prospect theory: Cumulative representation of uncertainty. *Journal of Risk and Uncertainty*, *5*, 297–323.

Karmarkar, U. S. (1978). Subjectively weighted utility: A descriptive extension of the expected utility model. *Organizational Behavior and Human Performance*, *21*, 61–72.

Karsh, R., Walrath, J. D., Swoboda, J. C., and Pillalamarri, K. (1995). *Effect of battlefield combat identification system information on target identification time and errors in a simulated tank engagement task*. (Technical report ARL-TR-854). Aberdeen Proving Ground, MD, United States: Army Research Laboratory.

Kogler, T. M. (2003). *The effects of degraded vision and automatic combat identification reliability on infantry friendly fire engagements.* Unpublished Masters thesis, Virginia Polytechnic Institute and State University, Blacksburg, VA.

Kramer, R. M. (1999). Trust and distrust in organizations: Emerging perspectives, enduring questions. *Annual Review of Psychology*, *50*, 569–598.

Lee, J. D. and Moray, N., (1992). Trust, control strategies and allocation of function in human–machine systems. *Ergonomics*, *35*, 1243–1270.

Lee, J. D. and See, K. A. (2004). Trust in automation: Designing for appropriate reliance. *Human Factors*, *46*, 50–80.

Lee, J. D. (2006). Human factors and ergonomics in automation design. In G. Salvendy (ed.), *Handbook of Human Factors and Ergonomics* (pp. 1570–1596). Hoboken, NJ: Wiley.

Lerch, J., Prietula, M., and Kulik, C. (1997). The Turing effect: The nature of trust in machine advice. In P. Feltovich, K. Ford and R. Hoffman (eds), *Expertise in Context: Human and Machine* (pp. 417–448). Cambridge, MA: AAAI/MIT Press.

Macmillan, N. A. and Creelman, C. D. (1991). *Detection theory: A user's guide.* New York: Cambridge University Press.

Madhavan, P. and Wiegmann, D. A. (2007). Similarities and differences between human–human and human–automation trust. *Theoretical Issues in Ergonomics Science*, *8*, 277–301.

Mayer, R. C., Davis, J. H., and Schoorman, F. D. (1995). An integrative model of organizational trust. *The Academy of Management Review*, *20*, 709–734.

Muir, B. M. and Moray, N. (1996). Trust in automation. Part II. Experimental studies of trust and human intervention in a process control simulation. *Ergonomics*, *39*, 429–461.

Neyedli, H. F. (2009). *Interface design for an automated combat identification system: Displaying reliability information.* Unpublished Masters Thesis, University of Toronto, Toronto, Ontario, Canada.

Neyedli, H. F., Hollands, J. G., and Jamieson, G. A. (2009). Human reliance on an automated combat identification system: Effects of display format. In *Proceedings of the Human Factors and Ergonomics Society 53rd Annual Meeting* (pp. 212–216). Santa Monica, CA: Human Factors and Ergonomics Society.

Parasuraman, R., Molloy, R., and Singh, I. (1993). Performance consequences of automation-induced 'complacency.' *International Journal of Aviation Psychology*, *3*, 1–23.

Parasuraman, R., Sheridan, T. B., and Wickens, C. D. (2000). A model for types and levels of human interaction with automation. *IEEE Transactions on Systems, Man and Cybernetics-Part A: Systems and Humans*, *30*, 286–297.

Prelec, D. (1998). The probability weighting function. *Econometrica*, *60*, 497–528.

Rasmussen, J., Pejtersen, A. M. and Goodstein, L. P. (1994). *Cognitive systems engineering.* New York: Wiley.

Rice, S., Clayton, K. D, and McCarley, J. (2009). The effects of automation bias on operator compliance and reliance. In D. H. Andrews, R. P. Herz, and M. B. Wolf (eds), *Human factors issues in combat identification* (pp. 263–274). Farnham, England: Ashgate.

Scerbo, M. W. (1996). Theoretical perspectives on adaptive automation. In R. Parasuraman and M. Mouloua (eds), *Automation and human performance: Theory and applications* (pp. 37–63). Mahwah, NJ: Lawrence Erlbaum Associates.

Sheridan, T. and Parasuraman, R. (2000). Human versus automation in responding to failures: An expected-value analysis. *Human Factors*, *42*, 403–407.

Sheridan, T. and Parasuraman, R. (2006). Human-automation interaction. *Reviews of Human Factors and Ergonomics*, *1*, 89–129.

Sherman, K. (2000). Combat identification system for the dismounted soldier. In *Proceedings of SPIE 2000: Digitization of the Battlespace V and Battlefield Biomedical Technologies II* (pp. 135–146). Orlando, FL: The International Society for Optical Engineering.

Sherman, K. B. (2002). Combat ID coming for individual soldiers. *Journal of Electronic Defense*, *25*, 34–35.

Sorkin, R. D. and Woods, D. D. (1985). Systems with human monitors: A signal detection analysis. *Human-Computer Interaction*, *1*, 49–75.

Spence, I. (1990). Visual psychophysics of simple graphical elements. *Journal of Experimental Psychology: Human Perception and Performance*, *16*, 683–692.

St. John, M., Smallman, H., Manes, D., Feher, B. A., and Morrison, J. G. (2005). Heuristic automation for decluttering tactical displays. *Human Factors*, *47*, 509–525.

Tatnall, A. (2000). Innovation and Change in the Information Systems Curriculum of an Australian University: A socio-technical perspective. PhD Dissertation, Faculty of Education. Central Queensland University.

Tversky, A. and Kahneman, D. (1992). Advances in prospect theory: Cumulative representation of uncertainty. *Journal of Risk and Uncertainty*, *5*, 297–323.

Von Neumann, J. and Morgenstern, O. (1944). *Theory of games and economic behavior*. Princeton, NJ: Princeton University Press.

Wang, L. (2007). *The effects of reliability information and interface features on trust and reliance on combat identification systems*. Unpublished Masters Thesis, University of Toronto, Toronto, Ontario, Canada.

Wang, L., Jamieson, G. A., and Hollands, J. G., (2008). Selecting methods for the analysis of reliance in automation. In *Proceedings of the Human Factors and Ergonomics Society – 52nd Annual Meeting* (pp. 287–291). Santa Monica, CA: Human Factors and Ergonomics Society.

Wang, L., Jamieson, G. A., and Hollands, J. G. (2009). Trust and reliance on an automated combat identification system *Human Factors*, *51*, 281–291.

Webber, S. S. (2002). Leadership and trust facilitation cross-functional team success. *Journal of Management Development*, *21*, 201–214.

Wickens, C. D. (2000). *Imperfect and unreliable automation and its implications for attention allocation, information access and situation awareness* (Technical Report ARL-00-10/NASA-00-2). Savoy, IL: University of Illinois Aviation Research Lab.

Wickens, C. D. and Hollands, J. G. (2000). *Engineering psychology and human performance* (3rd edn). Upper Saddle River, NJ: Prentice Hall.

Wickens, C. and Dixon, S. (2007). The benefits of imperfect diagnostic automation: A synthesis of the literature. *Theoretical Issues in Ergonomics Science. 8*, 201–212.

Wiegmann, D. A., Rich, A., and Zhang, H. (2001). Automated diagnostic aids: The effects of aid reliability on users' trust and reliance. *Theoretical Issues in Ergonomics Science*, *2*, 352–367.

Young, C. J. (2005). Fratricide. *(Dispatches: Lessons learned for soldiers)*, *11*(1). Kingston, Ontario: The Army Lessons Learned Center.

# Chapter 10
# User Trust in New Battle Management Technology: The Effect of Mistrust on Situation Awareness

Paul M. Salmon
*Monash University Accident Research Centre, Monash University, Victoria, Australia*

Neville A. Stanton
*School of Civil Engineering and the Environment, University of Southampton, Southampton, UK*

Guy Walker
*School of the Built Environment, Heriot Watt University, Edinburgh, UK*

Daniel Jenkins
*Sociotechnic Solutions, St Albans, UK*

Laura Rafferty and Kirsten Revell
*School of Civil Engineering and the Environment, University of Southampton, Southampton, UK*

## Introduction

Situation Awareness (SA) is now widely accepted to be a key consideration during system design (Endsley, Bolte and Jones, 2003; Salmon, Stanton, Walker and Jenkins, 2009). In the military, accurate SA is a critical commodity for successful operations; indicative of this is the fact that poor or diminished SA has been implicated in a number of fratricide or 'friendly fire' incidents occurring in the recent Iraq conflicts, including the Challenger II tank incident, the operation provide comfort Black Hawk incident (Gorman et al., 2006) and the recent Basra A-10 tank buster incident.

In recent times the military has moved towards the use of advanced technologies to enhance performance during warfare operations. Within the land warfare domain, one element of this has involved the development of digitised software-based battle management systems to support mission planning and battle execution activities. Proponents of these systems assert that, amongst other things, they will lead to significant enhancements in the levels of SA achieved by the teams using them (Alberts, Gartska and Stein, 1999; Meiter, 2006). Aside from the more obvious performance measures, one important consideration in the design and evaluation of these systems is the level of trust that users place in the SA-related information that is presented to them. Still relatively novel, the introduction of digitised systems typically replaces tried, tested and, more importantly, trusted methods of acquiring SA during mission planning and battle management. This chapter discusses the importance of user trust in new technology for SA acquisition and maintenance in the military. Using analysis of a recent digitised battle management system undertaken in the land warfare domain, the effect of poor design on user trust, and consequently the effect of user mistrust in technology on SA is discussed.

## Trust and Situation Awareness in Complex Collaborative Systems

### *Trust*

The analysis presented was undertaken in the context of land warfare mission planning and battle management activities, both of which involve distributed teams engaging in collaborative activities. Trust, both between team members and on behalf of the team members in the technology that they use, is critical to efficient distributed team performance. For example, mutual trust between team members has been identified as a key component of teamwork (e.g. Salas et al., 2005) and has been said to affect co-ordination, co-operation and helping within teams, as well as mediating the degree to which team members share ideas and opinions (Fiore et al., 2003). Accordingly, the level of trust between members of military teams has long been a key line of enquiry for researchers; for example, the level of trust in leaders (Sweeney, 2010), supervisors (Deluga, 1995) and other team members (Lee et al., 2010) have all been subject to investigation in the military context. Just as critical, but less well understood, however, is the level of trust that users place in work aids and technologies within military systems. Although trust in automation has received attention across the safety critical domains (e.g. Parasuraman and Wickens, 2008), it has not received significant attention in the military domain. In particular, how lack of trust influences important constructs such as SA remains largely unexplored in this context.

Trust in the technology being used, in particular with regard to the information being presented by technologies, is therefore of paramount importance in this domain. Despite this, little research has explored the links between trust and SA. Endsley and Jones (1997) suggest that confidence in data is a particular problem in combat environments since information is often dated, conflicting, interpreted incorrectly or patently false. They point out that 'the amount of confidence a crew member has in the accuracy and completeness of the information received and their higher level assessment of that information is a critical element of situation awareness' (Endsley and Jones, 1997, p. 28).

*Situation Awareness*

Various models of SA are presented within the academic literature, relating to the level of SA held by individuals (e.g. Endsley, 1995), teams (e.g. Salas et al., 1995) and sociotechnical systems (e.g. Salmon et al., 2009). For a detailed review and comparison of models the reader is referred to Salmon et al. (2008); however, for the purposes of this chapter it is worthwhile briefly discussing three distinct perspectives, each of which highlights the importance of trust in SA acquisition and maintenance. The individual psychological perspective argues that SA is purely an internal cognitive phenomenon and thus exists only in the working memory of human operators. This perspective is characterised by Endsley's three level model of SA (Endsley, 1995), which describes SA as a product, held in the minds of individual operators, that is distinct from the processes used to achieve it. Characterised as the individual's mental model of the current situation, SA comprises the following three levels: level 1, perception of the elements in the environment; level 2, comprehension of their meaning; and level 3, projection of future system states.

The so called computing/engineering perspective (Stanton et al., 2010), dominant in the military (Ackerman, 2005), argues that SA is situated in the world, whereby humans working in complex systems refer to the information held on displays and other artefacts as their 'SA'. This perspective highlights the importance of user trust in the 'SA' that is being presented to them by any work aid or technology, since designers often attempt to present SA information directly. For example, relating this approach to Endsley's model it is argued that directly presenting the higher levels of SA (e.g. level 2 and 3 SA-related information) on displays augments SA acquisition and maintenance since it removes the need to interpret level 1 data (Endsley et al., 2003). Mistrust in the information being presented by technology and displays under this viewpoint renders them useless in the context of SA acquisition and maintenance, since they attempt to provide SA directly to users.

Finally, the systems or Human Factors approach argues that SA is characterised by distributed cognition (Hutchins, 1995) and arises from the interaction between people and their environment; SA resides in the cycle of activity rather than any one operator alone and may be associated with agents but does not reside within them as it is borne out of the interactions between them. This viewpoint is characterised by Distributed Situation Awareness (DSA) models (e.g. Salmon et al., 2009) and is consistent with Human Factors/Ergonomics and a Socio-Technical Systems viewpoint (Walker et al., 2009). With regard to the analysis presented, the systems approach is of most relevance, since the level of SA held by the overall system, and the SA-related interactions between human and technological agents are of interest. From this viewpoint, user trust in technological agents is critical, since exchanges of awareness between humans and technology are key to collaborative system performance and SA.

### *Representing Situation Awareness*

To represent and evaluate SA, we use the propositional network methodology (Salmon et al., 2009). Based on the notion that knowledge comprises concepts and the relationships between them (e.g. Shadbolt and Burton, 1995), propositional networks provide a way of representing a system's SA and thus show the knowledge held by a system, the information underlying this knowledge and also the usage of this information by the different agents (human and non-human) involved. The propositional networks contain linked information elements representing the information used during task performance. These information elements represent what the agents 'needed to know' in order to successfully undertake task performance and are linked based on the causal links that emerged during the task under analysis (propositions such as <object> <relation> <subject> e.g. tank 'has' location, enemy 'has' plan etc). Constructing propositional networks involves collecting data regarding task performance through direct observation, cognitive task analysis and documentation review, and then identifying the information elements underpinning SA and the relationships between them by undertaking a simple content analysis on this input data. The key words related to task performance are extracted from the data and are then linked based on the causal links extant during the activities in question.

## Case Study: Digitised Battle Management System Evaluation

The technology that this analysis focussed on is a recently introduced digitised battle management system that provides support (i.e. planning tools and real-time situational display) for mission planning and execution tasks; the creators of the system claim that it has been designed in order to enhance SA, information management, planning, control and tempo during mission planning and execution. The system itself provides the user with various tools to support the planning process, such as mission analysis, decision support overlay matrix and synchronisation matrix construction tools. The battle management system also provides users with a real-time Local Operational Picture (LOP) display presenting friendly and enemy location and movement information and also various collaborative tools such as instant messaging and email facilities.

It is intended that the new system will replace the traditional mission planning and execution approach, labelled hereafter as the 'paper map' approach. Briefly, the paper map approach involved the development of planning products using paper maps, whiteboards, flipcharts and acetate overlays. Key elements related to the plan are drawn on acetate overlays (e.g. terrain analysis, commander's effects, situation overlays etc) and products are produced on paper, whiteboards (e.g. mission analysis) or on acetates (e.g. overlays such as the commander's effects schematic). During battle execution, enemy and friendly entities are placed on the maps using 'stickies' (plastic icons that can be stuck on and moved around the map).

The aim of the analysis presented was to assess the impact on SA that the new battle management system had during the mission planning and execution activities analysed. As part of an evaluation of the new digitised system, we used the propositional network approach to model SA during an operational field trial involving multiple mission planning and battle execution activities. The trial was set up specifically in order to test the new system and closely represented a real-world operational situation. The wider findings from the analysis are discussed in the context of DSA elsewhere (e.g. Salmon et al., 2009). For the purposes of this chapter a focus is placed on the level of trust placed in the system by users and the ensuing affect on SA acquisition and maintenance during planning and battle management operations.

## Methodology

### *Participants*

The study involved direct observation of mission planning and battle execution activities within the Brigade and Battle Group (BG) Headquarters (HQ). The participants involved in the study were the army staff working in the Brigade and BG teams. The Brigade and BG HQ consisted of the following cells for the staff to work from: G3 Operations, G5 Plans, G6 Operations, Combat Systems Support Operations (CSSO), Air Aviation, G2 Intelligence, ISTAR, I-Hub, Artillery and Engineers. Due to the nature of the study and data restrictions, it was not possible to collect participant demographic data.

### *Materials*

The materials used for this study included the digital battle management system terminals, the resources used by the Brigade and BG throughout the trial, including the materials at each HQ (e.g. maps, pens, VHF radios, acetates, tables, chairs, smart boards, standard operating instructions etc) and also on the battlefield (e.g. vehicles, equipment, weapons etc). The materials used by the analysts to collect the data included notepads and pens, digital cameras, and Dictaphones for audio recording purposes.

### *Procedure*

Six analysts undertook direct observation of activities in the Brigade and BG HQs. Analysts were located within the HQ during planning and battle execution and were given access to products, logs, briefs and subject matter experts throughout the field trial. Typically analysts directly observed activities during a particular mission planning cycle or execution of a battle. Upon completion of the planning/execution activities, analysts then held informal discussions with the key personnel involved in order to gather more detail regarding events of interest, to clarify unclear elements of the events observed, or to interrogate users of the battle management system regarding its usability and performance.

Propositional networks (Stanton et al., 2006; Salmon et al., 2009) were developed for the mission planning and battle execution activities observed based on verbal transcripts, task analysis and live observation. Due to the size of the experiment and breadth of the activities undertaken, the analysts focussed on small vignettes as opposed to entire scenarios; this approach is advocated by the NATO code of best practice (NATO, 2002) for studying command and control systems, suggesting that an appropriate way of dealing with the complexity of such systems is for analysts to present their findings in the form of vignettes.

## Results

As discussed, this chapter is focussed on levels of trust in new battle management system technology and its effect on SA. Therefore, we present here only extracts of results from the study described that relate to this issue. For the full SA analysis output, the reader is referred to Salmon et al. (2009).

### *Erroneous Information Presentation*

Key to augmenting user trust in new battle management system technologies is the presentation of accurate information during mission planning and battle execution. The propositional network analysis revealed various instances during mission planning and execution activities where the battle management system in fact presented erroneous information to its users. This was found to be problematic with regard to acquisition of SA, particularly during battle management activities. An example is presented in Figure 10.1, which is a propositional network derived from analysis of battle management activities during one of the battles observed. Within Figure 10.1, the nodes represent information elements that were 'active' during this phase of the battle, and thus being used by the Brigade members for SA acquisition and maintenance. Those information elements circled within the network represent information that was being presented inaccurately on the LOP at this point in the battle. Specifically, the exact location of a Javelin group on the battlefield, as presented in the Brigade HQ by the digital LOP, caused some uncertainty between the Chief of Staff (COS) and the Operations Officer (OpsO). An icon representing the current location of the Javelin group was presented on the LOP by the digitised system. Upon receipt of this information, the COS initially suggested that the Javelin group, as represented on the digitised system LOP, were in the wrong location (the LOP suggested that they were ahead of the recce group when, according to the mission plan they should have been behind them) and that they should be ordered to turn back and proceed to their correct position as required by the battle plan. Upon questioning the reliability of the LOP information presented by the battle management system, there was some debate between the COS and the OpsO as to whether the information regarding the Javelin group's location was correct, incorrect or out of date, following which the COS used radio voice communications to request confirmation of the Javelin group's position. After interaction with the Javelin group and the OpsO, the COS decided that the information presented on the LOP was incorrect and that the Javelin group were in fact in the appropriate location on the battlefield.

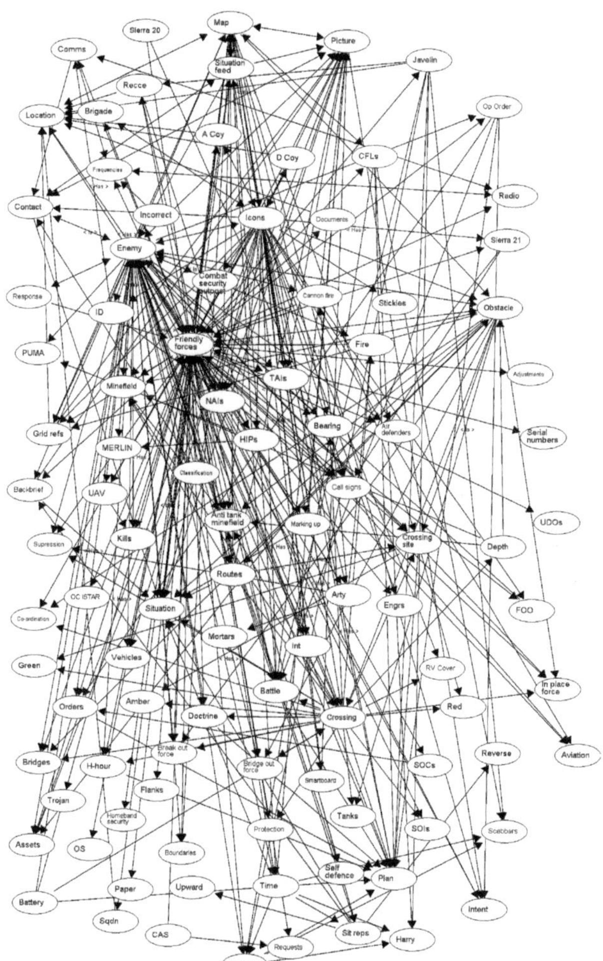


**Figure 10.1 Additional diagnostic decision requirement example; circled information elements show those involved in the diagnostic decision regarding the location of the Javelin group on the battlefield**

Various instances of the battle management system LOP display presenting erroneous information to users during battle execution activities were observed, and there were thus many instances in which users questioned the information being presented to them. As a corollary, users had to decide whether the information was correct, and if not, take further action to ascertain what the correct location and status of entities on the battlefield actually was. In most cases, user mistrust in the battle management system led them to take further actions to interrogate the information being presented, which added further processes to normal DSA development activities. A representation of the different SA development processes adopted when users trusted the information presented by the LOP versus instances when users did not trust the information is presented in Figure 10.2.

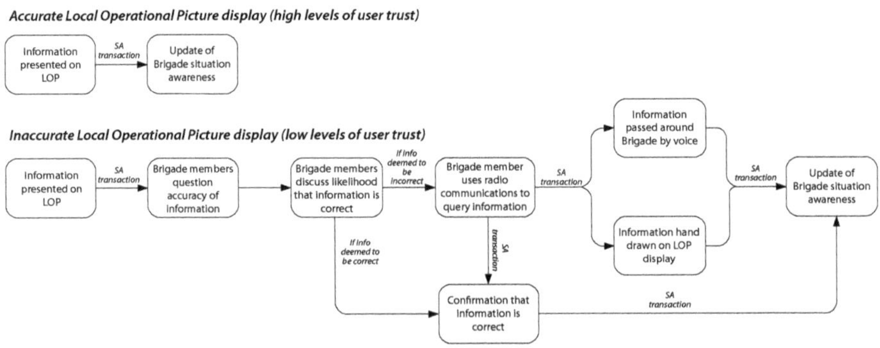


**Figure 10.2 Situation awareness development processes adopted with high levels of trust in the information presented by the battle management system LOP versus that adopted when users had low levels of trust in the information presented on the LOP**

As discussed, there were various instances in which information presented by the battle management system was found to be false, which, based on observations of planning and battle execution activities, and post-trial discussions with users, had the effect of reducing the level of trust that users placed in the new system. To summarise this, an overall propositional network was developed in order to emphasise those information elements, presented by the battle management system and underpinning SA acquisition and maintenance, that the users did not trust. This was developed based on a summary of the overall analysis, and is represented in Figure 10.3, whereby those information elements shaded darkly represent information elements presented by the battle management system that the users did not trust. These included the battlespace, friendly and enemy force movements, locations and updates, and mission planning products developed using the digitised system.

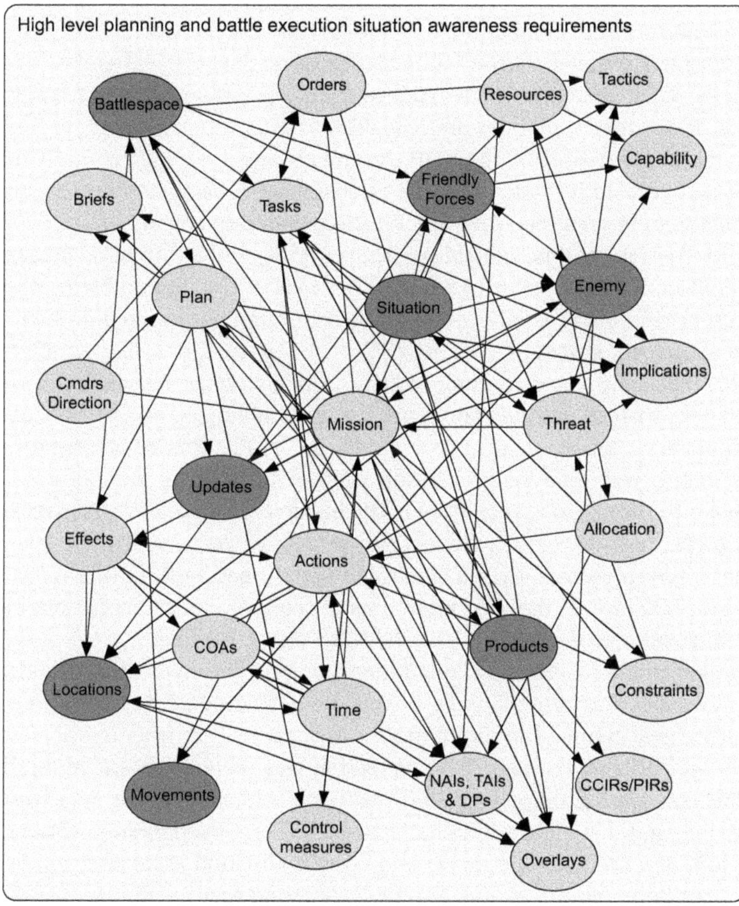


**Figure 10.3 Lack of trust in information elements; those information elements shaded in dark represent the information presented by the system that users felt was untrustworthy**

## Discussion

The aim of the analysis presented was to emphasise the important role that user trust in work technologies has in SA acquisition and maintenance. Unsurprisingly, based on the analysis presented, it is concluded that poor system design, such as allowing displays to present inaccurate, out-of-date information, has an adverse affect on the level of trust that users place in new work technologies. In this case, two key issues were identified that contributed to the low levels of trust placed in the battle management system by participants. First, there were many instances in which the information presented on the battle management system LOP was in fact inaccurate and was not compatible with the real state of the world at the time when it was presented. Often the information presented on the LOP was either out of date or spurious. This meant that the Brigade and BG's SA of enemy and friendly force locations, movements, number and capabilities was often inaccurate. Secondly, the information presented by the LOP was not always done so in a timely manner; due to data bandwidth limitations voice transmission was given precedence over global positioning data regarding the locations and movements of entities on the battlefield. Because of this, contact reports and positional information presented on the LOP was often up to 20 minutes late. Because of the two problems discussed above, users held a low level of trust in the information presented to them by the battle management system.

More importantly, and of more interest to this chapter, it was found that low levels of user trust in work technologies can also adversely affect SA acquisition and maintenance. In the analysis presented, the acquisition of SA regarding the ongoing battle was delayed as a result of user mistrust in the digitised battle management system. Low trust in the battle management system effectively added additional process to SA acquisition and maintenance. Users often questioned the information presented by the electronic system and took further measures to clarify the accuracy of the information (e.g. requests via radio for clarification of location and status reports). This had the effect of delaying SA acquisition and maintenance and reducing operational tempo; the level of SA held by the Brigade and BG was often lagging behind the real state of the world due to the added processes involved in clarifying the accuracy of information presented by the battle management system. SA acquisition was therefore made more difficult and more time consuming, since staff had to first recognise that data was likely to be wrong, and then take additional measures (i.e. voice communications using radios) to query the data and arrive at a correct situation assessment. It is worth pointing out here that the overall SA analysis found that the digitised system did not provide adequate support for SA development and maintenance during the planning and execution activities observed, with low user levels of trust in the system identified as one of the factors involved (Salmon et al., 2009).

Interestingly, user trust in the digitised battle management system was so low during the operational field trial that the Brigade and BG teams involved reverted back to the traditional paper map mission planning and battle management processes in order to support and supplement the new electronically supported process. On the majority of occasions this was because of flaws present within the new digitised system that were adversely impacting levels of SA. For example, throughout battle execution activities the Brigade team maintained a paper map representation of the battle, including real time locations of friendly and enemy units, in addition to the digitised LOP display battle representation. Information presented by the digitised LOP display was often cross-checked for accuracy with the paper map representation. When questioned why the paper map system was used, participants responded that it was quicker, easier and more effective to generate and maintain the level of SA required using the old paper map-based processes.

## Conclusion

It is concluded from the analysis presented that trust in work technologies is a critical factor in SA acquisition and maintenance. In particular, the analysis leads us to conclude that low levels of trust in technological work aids can adversely affect SA, primarily through delaying the acquisition and maintenance of SA through introducing the requirement for further interrogation of information for clarification purposes. It is therefore recommended that system designers consider the technological capability available and also strive to ensure the accuracy of all information presented by digitised systems. System designers need to carefully consider the constraints imposed on them by the technological capability available to them and design the system accordingly. User levels of trust in the case study presented were adversely affected because the capability of the displays used and data bandwidth limitations often rendered SA-related information out of date. It goes without saying that any information presented by command and control systems should be highly accurate. System designers need to ensure that the information presented by all aspects of the system is accurate at all times. The present study revealed that the battle management system did not always present accurate SA-related information, such as contact and positional reports, and enemy and friendly movements on the battlefield. This ultimately had the effect of lowering the levels of trust placed in the new system by its users.

In closing, it is clear that, due to technological advances, there is great potential to enhance levels of SA during military operations through the use of digitised mission support systems; however, low levels of trust in new technologies, brought about by poor system design and a lack of consideration of technological capability has the potential to adversely affect the level of SA afforded by such technologies. Further investigation and evaluation in order to determine how systems can be better designed in order to enhance trust, SA and ultimately mission performance, is required. Key issues to pursue relating to the concept of trust include how to maximise the accuracy of the information presented by such systems, and how to support the identification of erroneous information on real-time displays. Ultimately, the great potential that digitisation offers for enhancing command and control performance in the military domain is also accompanied by a very real opportunity to create untrustworthy warfare systems.

## Acknowledgement

The work reported in this chapter was undertaken as part of the Human Factors Integration Defence Technology Centre, which is part-funded by the Human

Sciences Domain of the UK Ministry of Defence Scientific Research Programme. Dr Paul Salmon's time in preparing this chapter was funded through the Australian National Health Medical Research Council post-doctoral training fellowship scheme. Finally, the authors are grateful to the many Armed Forces involved in the operational field trial who willingly participated in the study described.

## References

Ackerman, R. K. (2005). Army intelligence digitizes situational awareness. *Signal*, Retrieved November 19, 2008 from http://sncorp.com/.

Alberts, D. S., Garstka, J. J. and Stein F. P. (1999). *Network Centric Warfare*. Command and Control Research Program (CCRP), Washington DC.

Deluga, R. J. (1995). The relation between trust in the supervisor and subordinate organizational citizenship behaviour. *Military Psychology*, 7:1, pp. 1–16

Endsley, M. R. (1995). Towards a theory of Situation Awareness in dynamic systems. *Human Factors*, 37, pp. 32–64.

Endsley, M. R. and Jones, W. M. (1997). *Situation Awareness, Information Dominance, And Information Warfare*. Technical Report 97-01. Belmont, MA: Endsley Consulting.

Endsley, M. R., Bolte, B., and Jones, D. G. (2003). *Designing for Situation Awareness: An Approach to User-Centred Design*. London: Taylor and Francis.

Fiore, S. M., Salas, E., Cuevas, H. M. and Bowers, C. A. (2003). Distributed coordination space: toward a theory of distributed team process and performance. *Theoretical Issues in Ergonomics Science*, 4, pp. 340–364.

Gorman, J. C., Cooke, N., and Winner, J. L. (2006). Measuring team situation awareness in decentralized command and control environments. *Ergonomics*, 49, pp. 1312–1326.

Lee, A. Y., Bond, G. D., Russell, D. C., Tost, J., González, C., Scarbrough, P. S. (2010). Team perceived trustworthiness in a complex military peacekeeping simulation. *Military Psychology*, 22:3, pp. 237–261.

Hutchins, E. (1995). *Cognition in the Wild*. Cambridge, MA: MIT Press.

Meiter, J. S. (2006). Network enabled capability: a theory desperately in need of doctrine. *Defence Studies*, 6:2, pp. 189–214.

NATO (2002). *Code of Best Practice for C2 Assessment*. Department of Defense Command and Control Research Program (CCRP), 3rd edn.

Parasuraman, R. and Wickens, C. D. (2008). Humans: still vital after all these years of automation. *Human Factors*, 50:3, pp. 511–520.

Salas, E., Prince, C., Baker, P. D. and Shrestha, L. (1995). Situation awareness in team performance. *Human Factors*, 37, pp. 123–136.

Salas, E., Sims, D. E. and Burke, C. S. (2005). Is there a big five in teamwork? *Small Group Research*, 36:5, pp. 555–599.

Salmon, P. M, Stanton, N. A., Walker, G. H., Baber, C., Jenkins, D. P. and McMaster, R. (2008). What really is going on? Review of situation awareness models for individuals and teams. *Theoretical Issues in Ergonomics Science*, 9:4, pp. 297–323.

Salmon, P. M., Stanton, N. A., Walker, G. H., and Jenkins, D. P. (2009). *Distributed Situation Awareness: Advances in Theory, Measurement and Application to Teamwork.* Aldershot, UK: Ashgate.

Shadbolt, N. R., and Burton, M. (1995). Knowledge elicitation: A systemic approach. In J. R. Wilson and E. N. Corlett (eds) *Evaluation of Human Work: A Practical Ergonomics Methodology*, pp. 406–440.

Stanton, N. A., Stewart, R., Harris, D., Houghton, R. J., Baber, C., McMaster, R., Salmon, P. M., Hoyle, G., Walker, G. H., Young, M. S., Linsell, M., Dymott, R., and Green, D., (2006). Distributed situation awareness in dynamic systems: Theoretical development and application of an ergonomics methodology. *Ergonomics*, 49, pp. 1288–1311.

Stanton, N. A., Salmon, P. M., Walker, G. H. and Jenkins, D. P. (2010). Is situation awareness all in the mind? *Theoretical Issues in Ergonomics Science*, 11:1, pp. 29–40.

Sweeney, P. J. (2010). Do soldiers re-evaluate trust in their leaders prior to combat operations. *Military Psychology*, 22:1, pp. S70–S88.

Walker, G. H., Stanton, N. A., Salmon, P. M., and Jenkins, D. P. (2009). A review of sociotechnical systems theory: A classic concept for new command and control paradigms. *Theoretical Issues in Ergonomics Science*, 9:6, pp. 479–499.

# Chapter 11
# Active Trust Management

Elisabeth W. Fitzhugh
*SRA International, Inc., Dayton, OH, USA*

Robert R. Hoffman
*Institute for Human and Machine Cognition, Pensacola, FL, USA*

Janet E. Miller
*US Air Force Research Laboratory, Sensors Directorate, Dayton, OH, USA*

> The two truly transforming things, conceivably, might be in information technology and information operation and networking and connecting things in ways that they function totally differently than they had previously. And if that's possible, what I just said, that possibly the single-most transforming thing in our force will not be a weapon system, but a set of interconnections and a substantially enhanced capability because of that awareness (Secretary of Defense Donald H. Rumsfeld – Aug. 9, 2001, Town Hall Meeting, Washington, D.C.).

## Introduction

Much trust research focuses on the formation of trust as a singular state, and the topic of the human operator's trust in the automation. Discussions of trust by the Air Force Research Laboratory have converged on a notion of 'active trust management' (Hoffman et al., 2009). This concept of active trust management is complementary to the idea of trust formation. Why the focal shift?

In today's military environment, cyber technology is pervasive and cyber processes exist in a plethora of critical operations. No one can dispute the pervasiveness of electronic, networked systems within societies and the way increased dependence on these technologies has altered human activity. In 2001, then Secretary of Defense Donald Rumsfeld discussed the transformative power of networked automated electronic systems. The US Department of Defense (DoD) terms such computer networks as 'cyber systems' operating in 'cyberspace.' Although the rapid information exchange capabilities afforded by the move to cyberspace has strengthened all aspects of the social order, it has created vulnerabilities as well. Concern for the risk to national security posed by cyberspace threats and vulnerabilities led Secretary of Defense Robert Gates, in 2009, to create US Cyber Command (USCYBERCOM), a subordinate unified command under United States Strategic Command. US Cyber Command, developed in response to repeated foreign-based attacks on both Department of

Defense and US critical infrastructure networks, is tasked to coordinate computer-network defense and to direct US cyber attack operations (US Secretary of Defense, 2009).

DoD cyber systems are a form of military team. For example, Predators flying over Afghanistan are controlled by airmen somewhere in the United States; the communications that comprise flight control, sensor control, and sensor returns operate within an information network. The networks supporting Predator operations combine sensors, computers, satellite, radio, microwave, and landline communications infrastructure, and humans. Together, airmen in the US – directing Predator weapon systems operations – and computers – translating, transmitting and executing human direction – act as networked teams to accomplish military objectives. In *designing* military joint sociotechnical systems to operate in cyberspace, attention is paid to trust formation. System designers know the human team members must develop some level of trust in system components and the integrity of information exchanges in order to use the system. However, in *operating* military joint sociotechnical systems, real world experience demands the focus shift to trust management. An operational system in cyberspace, unlike a system on the design floor, can no longer be assumed to be under the operator's exclusive control. Trust management ensures that trust is merited. The watchword is former President Ronald Reagan's frequent quote, 'trust but verify.'

The defense of US cyber systems is challenging. Effects can be experienced in near real time from cyber actions taken halfway across the world. The actions may be temporally synchronized but they do not have to be spatially co-located. Because of the size and complexity of the networks employed, cyber operations are dependent upon an elaborate system of component subsystems and connections that can neither be continuously guarded nor fully secured. Researchers at the Tavistock Institute in the 1950s dubbed this type of system a 'sociotechnical work system' (see Hoffman and Militello, 2008; Vicente, 1999):

- *Large problem spaces* – Problems in these work systems have many attributes and, therefore, many possible solutions.
- *Social* – Tasks of this category involve more than one person working in a coordinated fashion.
- *Heterogeneous perspectives* – Although the high level goal for the participants might be the same, each of the participants has different concerns and viewpoints on reaching that goal. Oftentimes these differences cause conflicts which must be resolved for the effort to be satisfactorily completed.
- *Distributed* – The geographical separation can contribute to task difficulty because of human communication issues as well as time or cultural differences.
- *Dynamic* – The work methods are not static. Time dimensions can range from immediate responses to response times of months or even years.

- *Hazard* – Error can cause catastrophic results, be it health hazards, loss of life, financial ruin, or some other major complication.
- *Uncertainty* – While making decisions in sociotechnical fields is critical, the data input to the decision-making is often uncertain. Filtering through the input includes determining what is accurate, what is relevant and what is not represented.
- *Mediated interactions* – For many sociotechnical systems, what is really interesting and important cannot be directly observed.
- *Disturbances* – Complex systems often have unanticipated reactions and responses.

On this definition, DoD cyber systems certainly qualify as instances of sociotechnical work systems. Our focus in this chapter is on the fact that the subsystems – the connections, their functions, the data that are used and passed – all are subject to attack. The window of vulnerability for sociotechnical systems to be violated or corrupted never closes. System integrity must be continuously questioned and verified. In cyberspace operations, discussions of trust focus on a concept the Air Force Research Laboratory terms *active trust management* (Hoffman et al., 2009). The central idea is this: *Trust is not a state; it develops. It is dynamic, though there may be periods when it is relatively stable. Trust is not a single thing; it is multidimensional, contingent, and context dependent, even though there may be periods when it seems simple, fixed, and absolute.* To successfully leverage cyber capabilities, trust formation might profitably be rethought as active trust management. Much trust research focuses on trust formation in categorical terms (i.e., trust vs. distrust as distinct states). As might be expected, much of the language of trust research reflects that focus. Herein lies a fundamental distinction between research goals for *building* human–automation trust and those of *managing* trust among joint sociotechnical components. To maintain trust in that which is trustworthy is the goal in either case. However, in trust management in the cyber domain, trust remains conditional and trustworthiness might have to be reestablished with each transaction. Active trust management acknowledges that, in cyberspace as elsewhere, the price of freedom is eternal vigilance.

We call out the fact that much human factors research on trust focuses on the human operator's trust in automation. To some researchers, there is no significant distinction between interpersonal and human–computer interactions, viz., Johns' (1996) definition of a 'willingness to place oneself in a relationship that establishes or increases vulnerability with the reliance upon someone or something to perform as expected' (p. 81). Corritore et al. (2003) note two approaches to online truster/trustee relationships: individual-to-individual trust relationships mediated through technology vs. technology as the trust object itself. Such research demonstrates that people personify and relate personally to automation. This human tendency to attribute agency to automation (using terms such as benevolence, intent, and will) is both unavoidable and inappropriate. Woods (2009) and his colleagues (Hollnagel and Woods, 2006; Hollnagel et al., 2006) point out that, however autonomously they appear to operate, computers are not stakeholders in the processes in which they participate; they may perceive, but do not conceive, generative requirements. This is another reason for thinking of trust in terms of active management.

## Navigating a Morass of Concepts and Issues

Discussions of trust research can quickly become mired in the different perspectives from which researchers have approached trust phenomena: psychological, social, philosophical. A brief review is in order, to establish a common perspective for discussions of active trust management.

In its several semantic senses, trust can be associated with belief or state, with attitude, with intent, and with behavior. In search of conceptual clarity, Lee and See (2004) differentiate between the concepts of trust and reliance, following Fishbein and Ajzen's (1975) organizing framework. Fishbein and Ajzen reconcile conflicting definitions of trust by differentiating among trust behavior, intent, attitude, and belief. Within this simplistic description, behavior stems from intention, intention is a function of attitude, and attitude stems from belief/perception. While one may question the causal chain theory that Fishbein and Ajzen propose, all these factors are certainly in the mix for any descriptive theory of trust relations. In Fishbein and Ajzen's model, the trust belief is referred to as trustworthiness, the trust attitude is referred to as trustingness (willingness to trust), and the manifestation of trust in performance – the activity of trusting – is referred to as reliance.

We represent these concepts as a continuous cycle in Figure 11.1, to emphasize the analysis of trust in process terms (whether or not any simple cause–effect chains emerge in particular situations). Two things are especially noteworthy about the model expressed in Figure 11.1. First, there are no beginning and ending points. This reflects an empirical truth. Certainly, there can be apparent beginnings, as when two agents interact for the first time and a relation commences. Certainly, there can be stable periods in a trust relation, and certainly, there can be dissolutions to trust relations (as when one of the parties is no longer an agent in the relation). Second, Figure 11.1 hints at the complexity, interaction, and continuousness of the various factors. Trust is not simple.

This contrasts with the many definitions that approach trust as either a state or an attitude. The most widely accepted definition of trust is Mayer, Davis and Schoorman's (1995): 'a willingness to be vulnerable to another party when that party cannot be controlled or monitored.' This seems to identify trust as a state. Lee and See (2004), who examined trust from multiple perspectives, defined trust as 'the attitude that an agent [automated or human] will help achieve an individual's goals in a situation characterized by uncertainty and vulnerability.' If one is taking a purely *descriptive* perspective, Lee and See's definition seems appropriate to trust in sociotechnical systems. If, as in the case of active trust management, one has a *prescriptive* perspective, a static definition of trust, no matter how precise, is inadequate.

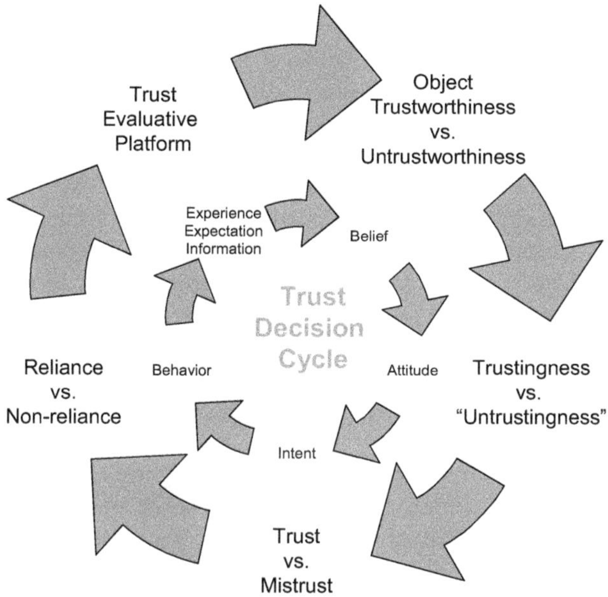


**Figure 11.1 A trust cycle, based on Fishbein and Ajzen's chain theory (1975)**

We assert that in active trust management, trust is most usefully conceptualized neither as a state to be achieved nor as an attitude, but as a macrocognitive process (Klein et al., 2003). We prefer to speak of *trusting*, rather than trust. We prefer to think of the various factors involved as being continuous, parallel, and highly interactive – even though at times there may be state-like stable points and emergent simple causal chains.

## A Framework for Active Trust Management

The process of trusting must be continuously managed and experimentally tested. Equally important, the process must be measurable. Examination of the trust process opens opportunities to measure trust development and maintenance in terms of rate, intensity, and probability of merit; trusting outcomes can be measured in terms of level of reliance. The trust experience (good or bad) can be examined to forecast probable future trust-related activity. Additionally, the effective conduct of operations in cyberspace demands that operationally relevant research encompass context. How, when, and why is trusting enhanced? How can it be reduced, and how can it be regained? The methods for active trust management must be integrated into information systems and networks at the systems-of-systems level.

Based on a study of aviation weather forecasting, Hoffman et al.(2006) proposed a simple matrix combining trust vs. mistrust and unjustified vs. unjustified. This is depicted in Figure 11.2. Forecasters were queried about their reliance on automation, and when asked, 'Do you trust the technology?' the senior forecasters typically answered, 'No.' Their response was immediate, and was dismissed with a chuckle. Neither the technology nor the data depicted are broadly trusted. In fact, the forecaster trusts that the data *will* be misleading, that the technology *will* crash, etc. Especially when coping with severe weather, verification was always sought by comparing data types and sources. What this suggested was that forecasters might be spending too much time living in a space characterized by unjustified mistrust, when they need support or information enabling them to shift to a space of justified trust and justified mistrust. Motion in the diagonals (Figure 11.2) would be the ideal trajectories for any active trust management system (Hoffman, 2009).

Although movement *can* go any way in the Figure 11.2 cells, the arrows indicate the preferred direction. In the military cyber domain, enduring constant hacker attacks from declared adversaries (enemy states and organizations) and opportunist adversaries (thrill-seeker hackers and cyber criminals), the challenges lie in the appropriate application of the trust. To move fearfully from justified trust to unjustified mistrust creates a capability deficit based on hesitance to rely on truly reliable systems provided to meet real needs. To move fearlessly from justified mistrust to unjustified trust creates a capability deficit based on naïve reliance on unreliable systems that either fail to perform or expose protected data.

The main point made by Figure 11.2 is that the single question, 'Do I trust this (software, machine, agent)?' can be profitably replaced by a host of questions, including:

'Under what circumstances do I trust?'
'To what degree do I trust?'
'Under what circumstances do I verify?'
'Under what circumstances do I mistrust?'

Some of this complexity is expanded in Tables 11.1, 11.2, and 11.3. We distinguish among three dyads: absolute vs. contingent trust, verified/justified vs. skeptical trust, and reliable vs. tentative trust (Hoffman, 2009; Hoffman et al., 2009). In this conception, trust phenomena are concerned with information validity, resource

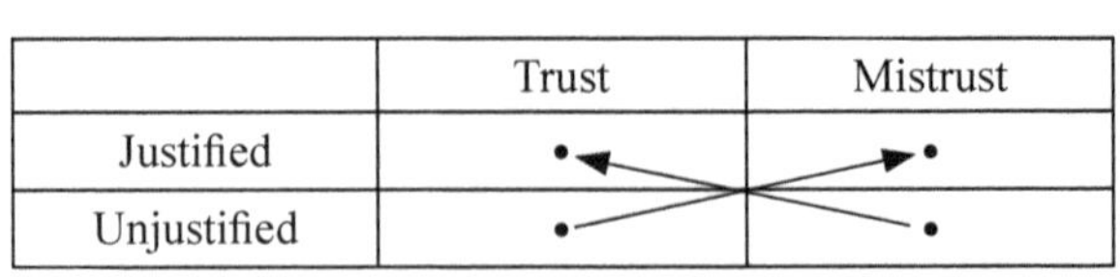


**Figure 11.2 An idealized trajectory model for trust dynamics**

management, and compliance. 'Can I trust that some questionable thing is true?', 'Can I trust resources are available/usable?', 'Can I trust that directives will be/can be/should be followed?' The first trust dyad, absolute vs. contingent trust, is predicated on the quality of the existence of trust. The second trust dyad, verified/justified vs. skeptical trust, is predicated on the quality of the rationale for trust. The third trust dyad, reliable vs. tentative trust, is predicated on the quality of the expression of trust.

**Table 11.1 Absolute vs. contingent trust (Trust is *when* ...)**

| **ABSOLUTE TRUST (Trust is *when* ...)** | | | |
|---|---|---|---|
| **TRUST IS ABOUT** ↓ | **Giving trust** | **Reciprocal** ← → | **Receiving trust** |
| | | | |
| **BELIEFS (data, information, knowledge)** | I can take X's assertions as true | ← → | X takes my assertions as true |
| **SURROGATES (resources, valuables)** | I can give X resources with directives as to their disposition and be certain that the resources will be disposed as directed | ← → | X can give me resources with directives as to their disposition and be certain that the resources will be disposed as directed |
| **ACTIONS (directives)** | I can give X directives and be certain that they will be carried out | ← → | X can give me directives and be certain that they will be carried out |
| **versus** | | | |
| **CONTINGENT TRUST (Trust is when ...)** | | | |
| **TRUST IS ABOUT** ↓ | **Giving trust** | **Reciprocal** ← → | **Receiving trust** |
| | | | |
| **BELIEFS (data, information, knowledge)** | I can take some of X's assertions as true under circumstances Z | ← → | X takes some of my assertions as true under circumstances Z |
| **SURROGATES (resources, valuables)** | I can give X resources with directives as to their disposition and be certain that X will try to dispose the resources as directed. | ← → | X can give me resources with directives as to their disposition and be certain that I will try to dispose the resources as directed. |
| **ACTIONS (directives)** | I can give X directives and be certain that X will try to carry them out. | ← → | X can give me directives and be certain that I will try to carry them out. |

**Table 11.2 Verified vs. skeptical trust (Trust is *why* …)**

| VERIFIED TRUST (or justified trust) (Trust is *why* …) | | | |
|---|---|---|---|
| **TRUST IS ABOUT** ↓ | **Giving trust** | **Reciprocal** ← → | **Receiving trust** |
| | | | |
| **BELIEFS (data, information, knowledge)** | I have reason to take X's assertions as true | ← → | X has reason to take my assertions as true |
| **SURROGATES (resources, valuables)** | I can give X resources with directives as to their disposition and have reason to be certain that the resources will be disposed as directed | ← → | X can give me resources with directives as to their disposition and have reason to be certain that the resources will be disposed as directed |
| **ACTIONS (directives)** | I can give X directives and have reason to be certain that they will be carried out | ← → | X can give me directives and have reason to be certain that they will be carried out |
| **versus** | | | |
| **SKEPTICAL TRUST (or mistrust) (Trust is *why* …)** | | | |
| **TRUST IS ABOUT** ↓ | **Giving trust** | **Reciprocal** ← → | **Receiving trust** |
| | | | |
| **BELIEFS (data, information, knowledge)** | I can take some of X's assertions as tentatively true | ← → | X takes some of my assertions as tentatively true |
| **SURROGATES (resources, valuables)** | I can give X resources with directives as to their disposition and be somewhat certain that X will try to dispose the resources as directed | ← → | X can give me resources with directives as to their disposition and be somewhat certain that I will try to dispose the resources as directed |
| **ACTIONS (directives)** | I can give X directives and be somewhat certain that X will try to carry them out | ← → | X can give me directives and be somewhat certain that I will try to carry them out |

**Table 11.3 Reliable vs. tentative trust (Trust is *how* …)**

| RELIABLE TRUST (Trust is *how* …) | | | |
|---|---|---|---|
| **TRUST IS ABOUT ↓** | **Giving trust** | **Reciprocal ← →** | **Receiving trust** |
| | | | |
| **BELIEFS (data, information, knowledge)** | I can take X's assertions as true most of the time | ← → | X takes my assertions as true most of the time |
| **SURROGATES (resources, valuables)** | I can give X resources with directives as to their disposition and be certain, most of the time, that the resources will be disposed as directed | ← → | X can give me resources with directives as to their disposition and be certain, most of the time, that the resources will be disposed as directed |
| **ACTIONS (directives)** | I can give X directives and be certain, most of the time, that they will be carried out | ← → | X can give me directives and be certain, most of the time, that they will be carried out |
| **versus** | | | |
| **TENTATIVE TRUST (Trust is *how*..)** | | | |
| **TRUST IS ABOUT ↓** | **Giving trust** | **Reciprocal ← →** | **Receiving trust** |
| | | | |
| **BELIEFS (data, information, knowledge)** | For the time being, I can take X's assertions as true | ← → | For the time being, X takes my assertions as true |
| **SURROGATES (resources, valuables)** | I can give X resources with directives as to their disposition, and for the time being, be certain that X will dispose the resources as directed | ← → | X can give me resources with directives as to their disposition, and for the time being, be certain that I will dispose the resources as directed |
| **ACTIONS (directives)** | I can give X directives, and for the time being, be certain X will carry them out | ← → | X can give me directives, and for the time being, be certain that I will carry them out |

The dyads in Tables 11.1–11.3 specify a both a challenge for cyberwork and a functionality for any active trust management system. The first member in each dyad represents an ideal; the second member in each dyad specifies the trust aspect that must be subject to active management. Given the persistence of cyber vulnerability:

- C*ontingent trust is preferable to absolute trust,* but how does one function using a system (or system-supplied data) that one is sure cannot be trusted? How can one continue to function, knowing confidently that the technology has been compromised?
- *Verified and skeptical trust are both desirable,* but it is not clear how to achieve the former or how to manage the latter. Skeptical trust is a pragmatic response to uncertainty, but again, how does one continue to function?
- *Reliable and tentative trust are both desirable.* While we know a great deal about evaluating the reliability of technology, is not clear how to actively manage the shifting states in which false alarm rates negatively impact trust in and reliance upon technology. Nor do we know how to depict context in order to show the operator the reasons why automation may be less trustworthy in given circumstances. Before cyber system deployment, designers strive to ensure that the technology merits reliable trust, but under threat of attack, fielded weapons systems or computer technologies may only merit tentative trust. Research on the human factors of trust in automation indicates that operators' willingness to use technology depends on whether the operator's trust is skeptical or tentative (Llinas et al., 1998).

The Air Force Scientific Advisory Board asked in its 2007 report on the Implications of Cyber Warfare, 'How does one fight through when trust is tentative at best and may even be totally unjustified?' The dyads of Tables 11.1–11.3 offer a window into opportunities for active trust management.

The fragmented and sometimes inconsistent approaches to trust definition, and consequently trust research, reflect a significant challenge to our conceptual understanding and operational definition of key terms. To advance on the path to implementing a scheme for active trust management, we need methods for measurement.

## Measurement Concepts

As noted above, trust itself has multiple definitions; however, certain components or subsidiary concepts are valid across different perspectives. Three concepts that merit consideration are: 1) Context of interest, 2) Measurable factors, and 3) Object of interest.

*Measurement Class: Context of Interest*

In order to observe and manage the trust process, it is important that the context of interest be sufficiently defined. For example, trust in the context of corporate financial dealings would be different from trust in the context of internet chat rooms. Also, various contexts can entail differing levels of various attributes, such as vulnerability, risk, and reward – all of which affect levels of trust. Levels of automation should also be included in the description of context and domain of interest. Key features of context include mode of information acquisition, mode of information analysis, process for decision and action selection, degree of automation, and method for implementation (Parasuraman et al., 2000).

*Measurement Class: Performance Measures of Interest*

As our discussion has shown, trust is an encompassing term that can and must be decomposed into measurable specifics that fit both the context of interest and the phase of the trust process addressed (Dekker and Hollnagel, 2004). For example, although loyalty was among the keywords identified by Adams et al. (2003) in their general literature review on trust, it is unlikely to be a useful trust construct in the context of human-to-machine and machine-to-machine interactions that characterize joint sociotechnical systems. Perkins et al. (2010) conducted an in-depth survey of literature on trust to compile a list of trust measures most frequently associated with trust in automation. Their list included the conceptual measurables presented in Table 11.4.

While competence, predictability, dependability, consistency, and confidence are commonly cited as measurable aspects of trust (in automation), recent efforts in several domains provide additional insight. Erickson (2009), in a US Air Force Research Laboratory Sensors Directorate effort to qualify trustworthiness in layered sensing, suggested security and safety as measurables. Schoorman et al., (2007), from a business perspective, identified integrity, ability, and benevolence as measurable attributes of human trust relations. Descriptions of resilience engineering for macrocognitive work systems have identified attributes of directability, responsiveness, reciprocity, and responsibility as measurables (Hoffman et al., 2009; Hollnagel et al., 2006). Any assessment of trust must specify which attributes of trust are logical measurable subcomponents for the context of interest, yielding potentially useful information when they are operationalized.

**Table 11.4 Some conceptual measurables for the evaluation of the dynamics of trust (after Perkins et al., 2010)**

| Conceptual measurable | Definition | Citations |
|---|---|---|
| | | |
| Competence | Competence is the ability to do the task at hand. The operator's perception of automation's competence is critical when making decisions. The operator must be able to judge the competency of the automation and use it appropriately | Many researchers cite competence as a major influential factor in trust formation and maintenance (e.g., Barber, 1983; Butler and Cantrell, 1984; Kee and Knox, 1970; Mishra, 1996; Lee and See, 2004) |
| Predictability | Predictability is the matching of performance with expectations. If the operator can predict what the automation should do, then the operator can understand when and why the technology fails, and can determine how to perform the task without the automation | Muir (1994) identified predictability as one of three critical contributors to a trust equation |
| Dependability | The operator must be able to rely on the automation to perform appropriately. If an operator cannot depend on the automation, then the automation is both pointless and useless | Noting the truster's vulnerability, a host of researchers assert that automation dependability is essential to trust (e.g., Lee and See, 2004; Mayer et al., 1995; Muir, 1994; Rempel et al., 1985) |
| Consistency | Automation's consistent performance (i.e., free from variation or contradiction) is essential to operator trust formation. If the automation fails to produce similar outcomes for identical tasks, the user's trust will be reduced. Inconsistency in automation is the first clue for a user to distrust and question the automation's validity. Consistency is the foundation of predictability | Many researchers regard consistency as one of the most influential factors (e.g., Butler and Cantrell 1984; Lee and See, 2004; Rempel et al., 1985) |
| Confidence | Confidence is a user feeling related to expectations of the automation; certainty that the automation will perform appropriately | Numerous trust researchers point out that the operator's sense of the automation's dependability and his/her confidence in the automation are crucial to a willingness to rely on system (e.g., Lee and See, 2004; Moray et al., 2000) |

*Measurement Class: Object of Interest*

As our discussion has made clear, the evaluation of trust differs significantly comparing trust in automation with interpersonal trust. Any scheme for active trust management will need a robust capability to describe various 'objects of interest' in trust measurement. Research indicates the importance of the object of trust and the ability of individuals to discriminate trustworthiness of different targets (Madhavan and Wiegmann, 2007). Just as love is a relationship with a lover and an object of love, trust is a relationship with a truster and object of trust. And just as with love, the relationship may or may not be reciprocal. The object of trust is context dependent. In the case of joint sociotechnical systems, potential objects include the operators who interact with the system, the automated system itself, the network over which information is routed, and the information that is passed. Each object of trust requires a different framework for operationalizing useful measures.

## Focus Areas for the Active Management of Trust

The approach proposed by the Trust in Cyberdomains Working Meeting defined four focus areas for active trust management (Hoffman, Patterson and Miller, in press). Each focus area highlights critical trust issues and challenges. The four focus areas are: trust in technology, trust in decision-making, trust in the sociotechnical work system, and trust in overall capability.

The four focus areas of trust can support analysis of trust phenomena in military teams and provide a frame from which a scheme for active trust management may be implemented and extrapolated. Although the focus areas share common themes (e.g., vulnerability, risk, uncertainty, reliability, integrity), the focus of trust is distinctive. We will now look at each facet specifically in the context of managing trust in military teams. We will use the Predator weapon system as an example to concretize the ideas.

*Trust in Technology*

Trust in technology encompasses issues and challenges stemming from the perceived integrity that is attributed to technology, whether justified or unjustified. Issues include considerations of the security and reliability of the technology, data integrity, transmission integrity, establishment of source credibility (certificate exchanges), system opacity, and system reliability.

Often, technology developers assume that operators will trust the technologies they develop. Certainly, the developers of military cyber systems expect that operators will both trust and employ their systems. However, operators are often (if not typically) very cautious and mistrusting, until they learn the competence envelope of the technology. This reticence is justified. Trust in technology should not be assumed and should always be questioned. At any given time, trust may be justified or unjustified, and conversely, mistrust may be justified or unjustified. Too little trust placed in the automation can lead to skepticism and underuse or disuse. Too much trust can lead to complacency and overuse or misuse.

Current thought is that these tendencies are connected to the nature of human reasoning and emotion, which is predicated on experiences with other humans. Research suggests that the attribution of integrity to automated systems is often considered analogous to the attribution of integrity among humans. People sometimes anthropomorphize the technology, relating to it as if it were human (Dzindolet et al., 2003). The key to improving human–automation collaboration is to define the boundaries for anthropomorphic stances and the competence envelopes of automated systems, and to identify and leverage distinctions in how the user should build (and delimit) trust in automation vs. in humans.

But where and under what circumstances does the human need to develop trust and where to maintain skepticism? For the military team to employ appropriate trust, the components that are vulnerable to compromise must be identified, and protective security and security monitoring measures must be in place and observable by the operator of the technology. For the team to maintain appropriate trust, trustworthiness must be observable, security conscious and security co-operative activities must be employed, and security activities verified.

Protective security is the organized system of defensive measures instituted and maintained at all levels of command with the aim of achieving and maintaining security (Joint Chiefs of Staff, 2005). From a cyber system perspective, the ideal protective security would shield system data and system function. However, we know that this is not uniformly possible. The game of 'catch-up' with clever hackers is a never-ending game. Protection of data and transmission integrity and the establishment of source credibility (e.g., web service trust measures, certificate exchanges, cyber identities) are cyber security technology challenges.

Another objective is system opacity, here used to mean specifically that the working of the automation is shielded from external observation but is observable to the operators of the technology. A protective method, security monitoring, is 'the collection, analysis and escalation of indications and warnings to detect and respond to intrusions' (Bejtlich, 2004). Rule-based alerts and software mechanisms that isolate suspected problem areas or suspend compromised functions are believed necessary to maintain security. When the operators are apprised of and respond appropriately to overt intrusion schemes and suspicious network symptoms, they, too, become part of system security – *if* they respond appropriately. Even small changes in the false alarm rate of protective measures (e.g., alerts) can drastically decrease trust in and reliance on technology, even if hit rates remain high (Lee and See, 2004).

Another significant consideration for trust in automation is that the technology might be compromised in a manner that is intentionally obscured. While operators may recognize signs of a virus-laden system (slow performance, full hard drive, etc.), a well-designed covert intrusion does not exhibit such signs. Thus, a subtle and dangerous situation develops when operators continue to trust systems that are no longer trustworthy. This reinforces the claim that trust should be thought of as a thing to be actively managed: constantly, deliberately, and actively monitored. In terms of military team performance, this means institution of checks, even in the absence of indicators and warnings.

### *Trust in Decision-making*

This encompasses trust-related issues and challenges stemming from vulnerabilities inherent in transmission of decision information among humans or humans and machine agents. The object of analysis is the decision-maker's ability to assess decision information reliability and to deal with uncertainty. Example issues include considerations of compromised decision information in storage, transmission, or processing.

Decision-making also involves the context of the situation and the relevant components of trust for the context. These are factors to which computers are essentially blind. The object of trust is the decision-maker's ability to discriminate on issues such as knowing how much uncertainty exists and the appropriate extent of trust. Decision-making is impacted by other agents' reasoning and intent, which may be known only to themselves. This is compounded by the fact that information about another agent's reasoning and intent is itself mediated, in part, by information that is provided by technology. These factors may give rise to forms of uncertainty and risk. This is apparent in the Mayer et al., definition of trust:

> The willingness of a party to be vulnerable to the actions of another party based on the expectation that the other will perform a particular action important to the truster, irrespective of the ability to monitor or control that other party. This definition of trust is applicable to relationship with another identifiable party that is perceived to act and react with volition towards the truster (Mayer et al., 1995, p. 712).

The potential risk incursion gives rise to a *trust dilemma* (Kramer, 2006). Kramer describes trust dilemmas as a kind of 'Catch 22' situation:

> In a trust dilemma, decision-makers hope to reap some perceived benefits from engaging in trusting behavior with another (agent). Pursuit of that opportunity, however, exposes them to the prospect that their trust may be exploited or betrayed. This conjunction of opportunity and vulnerability is, of course, the *sine qua non* of a trust dilemma (p. 68).

### *Trust in the Sociotechnical Work System*

This encompasses trust-related issues and challenges stemming from the perceived and real vulnerabilities and the risk tolerances that characterize interactions among the agents that constitute the broader sociotechnical work system, spanning teams and the entire organization. The object of analysis of trust in the sociotechnical work system includes risk and uncertainty tolerance, performance predictability/reliability, and agent accessibility and accountability.

Different domains differentially emphasize the characteristics of sociotechnical work systems. The goal in all sociotechnical work systems is to optimize the human, the technology's abilities and their interrelationship. In sociotechnical systems, trust is established by experience, but also by the agreement upon rules and adherence to rule-governed activities (Klein et al., 2004). The automated portions of the sociotechnical system are not self-aware partners and have no free will (Hollnagel and Woods, 2006) – they are not stakeholders. In contrast to Mayer et al.'s definition, trust in system components that can be breached by external, adversarial interests *must* be conditional upon the ability to monitor and direct the other party. One cannot assume goodwill on the part of an automated system, no matter how often it performs to specification.

One key to active trust management is vigilance in security monitoring. However, vigilance, especially under either conditions of high workload or stretches of unrelieved tedium, is difficult to maintain. Vigilance issues, in terms of both the behavior of the technology (monitoring) and continuous employment of security-conscious (protective) activities, are challenges for active trust management. Alarms and alerts might ameliorate human operator vigilance vulnerabilities, and potentially, enhance the trustworthiness of the weapon system. However, the challenge does not disappear with the addition of more and more technological fixes. Rather the work changes; in this case, the human now has the challenge of monitoring the intrusion detection system and maintaining an appropriate level of trust in that new layer of technology.

### *Trust in Overall Capability*

A distinction can be drawn between trust in the sociotechnical system (i.e., the technology and the workers) vs. trust in the overall capability of the system to achieve its primary task goals. Trust in capability refers to confidence that the total work system, from the level of individual teams up through the entire organization, can competently and safely achieve its principal task goals in context. The operators need to maintain confidence that 'we are getting the job done.' A sociotechnical work system may perform competently in one arena, but may not when transitioned to a context for which it was not originally intended and in which it may prove brittle (or non-resilient). The issue of reuse simply to save dollars is one impetus for this situation; the current DoD-mandated pursuit of net-centric data goals is another. The object of trust, then, is on the output performance of the work system in context. Issues in addition to raw performance are entrained, such as perceived utility (i.e., usefulness, completeness, and timeliness).

We have emphasized the application of the active management notion to the cyber arena, but it applies to all weapon systems.

## Application Example: The Predator Weapon System

Table 11.5 illustrates how the four focus areas might manifest in respect to the Predator weapon system for the Performance of Interest and Object of Interest measurement classes. (Table five does not list the Context of Interest variables since those would be shared for all the focus areas, and would be couched in terms of mission phases: take-off/land, taxi, navigate, loiter, surveil). Trust evaluation would include understanding the information flow between the facets in the networked system (perhaps using Figure 11.2 as an extremely high level view.) Basically, this would be a systems analysis but with particular emphasis on the concept of trust.

In this weapon system context, the active management of trust implies that continuously shifting states of trust must be observable as the work system is in action. Trust measurement must enable observation of the level of trust a human has for an agent, perhaps by having the user annotate trust levels in a log or using a radar or other diagram, such as is discussed in Miller and Perkins (2010). Active monitoring could be implemented under information assurance programs within DoD. Operators would need to be educated to understand that trust is an important part of vulnerability and that vulnerability to threats must be continuously evaluated and experimented.

**Table 11.5 Interpreting the focus areas for active trust management with respect to the Predator weapon system**

| TRUST IN TECHNOLOGY | | |
|---|---|---|
| | **Measurement class: Object of interest** | **Measurement class: Performance of interest** |
| Trust in the technology would require controllers at distant centers to verify that their control signals were being properly executed and that what they see on the screen is, in fact, what is being sensed by the Predator. Additionally, imagery and data analysts reviewing the resulting data feeds must ascertain whether to trust that the data received has not been tampered with or otherwise lost integrity | Predator aircraft<br>Sensors<br>Monitoring devices | Competence<br>Predictability<br>Dependability<br>Consistency<br>Confidence |
| **TRUST IN DECISION-MAKING** | | |
| In order to function as an effective decision aid, the system integrity must have embedded security checks and intrusion warnings, it must make the boundaries of its competence envelope continually observable, and the technology function must be monitored for anomalies | Humans-in/on-the-loop<br>Sensors | Integrity<br>Ability<br>Benevolence |
| **TRUST IN THE SOCIOTECHNICAL WORK SYSTEM** | | |
| Task-weary operators may benefit from automated system intrusion detection aids that not only alert the operator but also dispatch automated systems or joint teams to resolve detected anomalies. In an article on an automated security system, Li and Parker (2008) discuss a robot-aided intrusion detection system exhibiting high reliability detecting and resolving time-related anomalies | Complete work system, spanning operations, effects, teams, organizations. | Directability<br>Responsiveness<br>Reciprocality<br>Responsibility |
| **TRUST IN CAPABILITY** | | |
| For the Predator weapon system, the sensors have a specific field of view designed to meet current needs. If repurposed, each component within the system might itself remain trustworthy. However, trust in its overall capability might be diminished as its field of view caused its sensor returns to be less interpretable | Communications<br>Primary mission objectives | Security<br>Safety<br>Reliability |

## Conclusion

> Much of the trust research is about trusting information technologies that may be imperfect, or brittle and limited. That orientation is different from cyber warfare, where we will be dealing with an intelligent and deceptive adversary. It is not a question of whether our systems will be corrupted. They are going to be corrupted. They will be corrupted in ways that may quickly be obvious, such as denial of service. They will be corrupted in ways that we won't easily detect – our adversaries will be deceptive. It is better to have us trust compromised systems than stop using them. As a result, the traditional research on trust in information technology may be beside the point (Klein, 2009).

Sociotechnical work systems do not function exactly as human teams because the automated component has no loyalty or conscience, and it forms no emotional bonds. It can be breached and Compromised, and it is the responsibility of the human team member to remain vigilant and alert to loss of system trustworthiness. At the same time, the technology and the work methods it shapes are stand-ins for the intent of the designers and the mental model the designers have of the work that is to be performed. These complexities and subtleties are critical considerations in the development of active trust management research concepts and schemes for military work. The goal of active trust management is not simply trust formation, but rather continuous monitoring to manage a system in which one maintains not absolute, but appropriate trust. Additionally, the trust must be constantly tested because, just as one country is trying to secure its capabilities so that the task at hand can be performed, other forces are as diligently trying to leverage vulnerabilities, or such as conducting cyber warfare, so that the efforts are thwarted. Active trust management requires a commitment to constant vigilance and the constraint to trust but verify. Viewing trust formation as a continuous process (trusting) and active trust management as a dynamic, responsive, ongoing effort will be vital to shaping meaningful research strategies. Trust establishment is iterative, with each experience feeding into or detracting from future trust attribution.

One of the most challenging, yet unavoidable implications of our analysis (and the concept of 'fighting through') is that the active management of trust must include active experimentation on trust *in the operational context.* This will be a radical departure from traditional military methods, in which research is strictly firewalled from operations. But the need to include experimentation in the operational context is being forced upon us by the new complications wrought by the cyberworld.

**References**

Adams, B., Bruyn, L. Houde, S., and Angelopoulos, P. (2003). *Trust in automated systems literature review*. Report No. CR-2003-096. Toronto, Canada: Defence Research and Development.

Air Force Scientific Advisory Board (August 2007). *Report on implications of cyber warfare, volume 1: Executive summary and annotated brief.* Report No. SAB-TR-07-02. Washington, DC: US Air Force.

Barber, B. (1983). *The logic and limits of trust.* New Brunswick, NJ: Rutgers University Press.

Bejtlich, R. (2004). *The Tao of network security monitoring: Beyond intrusion detection.* Boston: Addison Wesley Professional.

Butler, J. K. and Cantrell, R. S. (1984). A behavioral decision theory approach to modeling trust in superiors and subordinates. *Psychological Reports, 55,* 19–28.

Corritore, C., Krachera, B., and Wiedenbeck, S. (2003). On-line trust: Concepts, evolving themes, a model. *International Journal of Human-Computer Studies, 58,* 737–758.

Dekker, S. W. A. and Hollnagel, E. (2004). Human factors and folk models. *Cognition, Technology and Work, 6,* 79–86.

Dzindolet, M. T., Peterson, S. A., Pomranky, R. A., Pierce, L. G., and Beck, H. P. (2003). The role of trust in automation reliance. *International Journal of Human-Computer Studies, 58,* 697–718.

Erickson, J. (2009). Trust metrics. In *Proceedings of 2009 International Symposium on Collaborative Technology and Systems,* 18–22 May 2009, Baltimore, Maryland.

Fishbein, M. and Ajzen, I. (1975). *Belief, attitude, intention, and behavior: An introduction to theory and research.* Reading, MA: Addison-Wesley.

Hoffman, R. R., Coffey, J. W., Ford, K. M., and Novak, J. D. (2006). A method for eliciting, preserving, and sharing the knowledge of forecasters. *Weather and Forecasting, 21,* 416–428.

Hoffman, R. R. and Militello, L. G. (2008). *Perspectives on cognitive task analysis: Historical origins and modern communities of practice.* Boca Raton, FL: CRC Press/Taylor and Francis.

Hoffman, R. R. (March 2009). A factor analysis for trust. [Working Paper.] Pensacola Beach, FL: Institute for Human-Machine Cognition (IHMC). Work done for 711th Human Performance Wing, Human Effectiveness Directorate, under USAF Contract Number: FA8650-04-D-6405.

Hoffman, R. R., Lee, J., Woods, D., Shadbolt, N., Miller, J., and Bradshaw J. (2009). The dynamics of trust in cyber-domains. *IEEE Intelligent Systems, 24*(6), 5–11.

Hoffman, R. R., Patterson, E. and Miller, J. (in press). *Trust in cyberdomains research roadmap recommendations: Findings from a working meeting, 14–*

*16 July 2009 Pensacola, FL*. Wright-Patterson AFB, OH: Air Force Research Laboratory, Human Effectiveness Directorate.

Hollnagel, E. and Woods, D. D. (2006). *Joint cognitive systems: Foundations of cognitive systems engineering*. Boca Raton, FL: Taylor and Francis.

Hollnagel, E., Woods, D. D., and Leveson, N. (eds) (2006). *Resilience engineering: Concepts and precepts*. Hampshire, UK: Ashgate.

Johns, J. (1996). A concept analysis of trust. *Journal of Advanced Nursing, 24*, 76–83.

Joint Chiefs of Staff. (2005). *Joint Publication 1-02, Dictionary of Military and Associated Terms*. Washington, DC: US Department of Defense, Joint Chiefs of Staff.

Kee, H. W. and Knox, R. E. (1970). Conceptual and methodological considerations in the study of trust and suspicion. *Conflict Resolution, 14*, 357–366.

Klein, G., Ross, K. G., Moon, B. M., Klein, D. E., Hoffman, R. R., and Hollnagel, E. (May/June 2003). Macrocognition. *IEEE: Intelligent Systems*, 81–85.

Klein, G., Woods, D. D., Bradshaw, J. D., Hoffman, R. R., and Feltovich, P. J. (November/December 2004). Ten challenges for making automation a 'team player' in joint human-agent activity. *IEEE: Intelligent Systems*, 91–95.

Klein, G. (July, 2009). 'Infirmation' Technology. [Position paper.] In R. Hoffman, E. Patterson, and J. Miller (in press), *Trust in cyberdomains research roadmap recommendations: Findings from a working meeting, 14–16 July 2009 Pensacola, FL*, pp. B26–27. Wright-Patterson AFB, OH: Air Force Research Laboratory, Human Effectiveness Directorate.

Kramer, R. (2006). Trust as situated cognition: an ecological perspective on trust decisions. In R Bachmann and A. Zaheer (eds), *Handbook of trust research.* Northampton, MA: Edward Elgar.

Lee, J. D. and See, K. A. (2004). Trust in automation: Designing for appropriate reliance. *Human Factors, 46*, 50–80.

Li, Y. and Parker, L. (2008). Detecting and monitoring time-related abnormal events using a wireless sensor network and mobile robot. *Proceedings of IEEE International Conference on Intelligent Robots and Systems*, Nice, France.

Llinas, J., Bisantz, A., Drury, C., Seong, Y., and Jian, J. (1998). *Studies and analyses of aided adversarial decision making. Phase 2: Research on human trust in automation.* Report No. AFRL-HE-TR-1999-0216. Wright-Patterson AFB, OH: Air Force Research Laboratory, Human Effectiveness Directorate.

Madhavan, P. and Wiegmann, D. (2007). Similarities and differences between human-human and human-automation trust: An integrative review. *Theoretical Issues in Ergonomics Science, 8*(4), 277–301.

Mayer R. C., Davis J. H., and Schoorman, F. D. (1995). An integrative model of organizational trust. *Academy of Management Review, 20*(3), 709–734.

Miller, J. and Perkins, L. (2010). 'Development of Metrics for Trust in Automation.' In Proceedings of the 15th International Command and Control Research and Technology Symposium, 22–24 June 2010, Santa Monica, CA.

Mishra, A. K. (1996). Organizational response to crisis. In R. M. Kramer and T. R. Tyler (eds), *Trust in organizations: Frontier of theory and research* (pp. 261–287). Thousand Oaks, CA: Sage.

Moray, N., Inagaki, T., and Itoh, M. (2000). Adaptive automation, trust, and self-confidence in fault management of time-critical tasks. *Journal of Experimental Psychology: Applied*, *6*, 44–58.

Muir, B. M. (1994). Trust in automation: 1. Theoretical issues in the study of trust and human intervention in automated systems. *Ergonomics*, *37*, 1905–1922.

Parasuraman, R., Sheridan, T. B., and Wickens, C. D. (2000). A model of types and levels of human interaction with automation. *IEEE Transactions on Systems, Man, and Cybernetics – Part A: Systems and Humans*, *30*, 286–297.

Perkins, L., Miller, J. Hashemi, A. and Burns, G. (2010). Designing for human-centered systems: Situational risk as a factor of trust in automation. *Proceedings of the 54th Annual Meeting of the Human Factors and Ergonomics Society*, 27 September–1 October 2010, San Francisco, CA.

Rempel, J. K., Holmes, J. G., and Zanna, M. P. (1985). Trust in close relationships. *Journal of Personality and Social Psychology*, *49*, 95–112.

Rice, S. and Keller, D. (2009). Automation reliance under time pressure. *Cognitive Technology*, *14*(1), 36–44.

Rumsfeld, D. H. (2001). Pentagon Town Hall Meeting, August, 9, 2001. Washington, DC: US Secretary of Defense. [Speech.] Accessed online 29 April 2010 at http://www.defense.gov/speeches/speech.aspx?speechid=1101.

Schoorman, F. D., Mayer, R. C., and Davis, J. H. (2007). An integrative model of organizational trust: Past, present, and future. *Academy of Management Review*, *32*(2), 344–354.

US Secretary of Defense. (2009). Establishment of a Subordinate Unified U.S. Cyber Command under U.S. Strategic Command for Military Cyberspace Operations. [Memorandum.] Accessed online 29 April 2010 at http://www.govexec.com/nextgov/0609/gates_cybercommand_memo.pdf.

Vicente, K. J. (1999). *Cognitive work analysis: Towards safe, productive, and healthy computer-based work*. Mumaw, NJ: Lawrence Erlbaum.

Woods, D. (2009). *Trust emerges from the dynamics of reciprocity, responsibility and resilience in networked systems.* [Briefing.] Presentation at the Air Force Research Laboratory Human Effectiveness Directorate's Trust in Cyberdomains Working Meeting, Pensacola, FL, 14–16 July 2009.

# Chapter 12
# Teamwork and Trust: A Socio-Technical Perspective

A. J. Masys
*University of Leicester, Leicester, UK*

## Introduction

Military operations are characterized by interdependent teams working in an environment of high risk, high tempo, and complexity. Teamwork and trust is a requirement in military operations because of the need to depend on the performance of others in the conduct of interdependent tasks. Increasing technological demands associated with military operations, such as in aviation, requires understanding regarding the role of trust in mediating human–automation interaction. Teamwork thereby emerges within a socio-technical network of elements whereby trust shapes communication processes and enables cooperative work team tasks.

Sociology offers an interesting approach for looking at the socio-technical elements of complex systems through the application of actor network theory (ANT). The systems perspective of ANT looks at the inter-connectedness of the heterogeneous elements characterized by the technological and non-technological (human, social, organizational) elements. The network space of the actor network provides the domain of analysis that presents teamwork and trust as a network of heterogeneous elements that shape and are shaped by the network space. Yeung (2002) notes that much of the work that draws on actor network theory places its analytical focus on unearthing the complex web of relations between humans and non-humans. The interaction of non-human actors with the human actors (such as a pilot) gives shape and definition to identity and action. Latour (1994b:806) argues that 'it is impossible even to conceive of an artifact that does not incorporate social relations, or to define a social structure without the integration of non-humans into it'. Leveraging the ANT perspective that every human interaction is socio-technical, teamwork and trust within the context of military operations is described as 'materially heterogeneous' (Callon and Law, 1997:166).

This chapter, derived from the work of Masys (2010), explores teamwork and trust in military operations through the lens of ANT, thereby revealing the socio-technical perspective.

### *System Thinking*

System thinking is both a worldview and a process in the sense that it informs one's understanding regarding a system and can be used as an approach in problem solving (Edson, 2008:5). 'Systems theory' represents a theoretical framework, a perspective and a set of methodological tools that may be applied to any field of study. The systems perspective reveals properties of the whole that are not evident with an examination of the components, thereby revealing emergent behaviour that arises from the dynamic interaction of components. Germane to this work, the socio-technical system is a topic of inquiry within sociology that combines the social and technical paradigms and examines the relationship between them. As described by Coakes (2003:2), 'Socio-technical thinking is holistic in its essence; it is not the dichotomy implied by the name; it is an intertwining of human, organizational, technical and other facets'. Senge (1990) argues that since the world exhibits qualities of wholeness, the relevance of systemic thinking is captured within its paradigm of interdependency, complexity and wholeness. The systemic perspective of ANT makes it well suited to facilitate an examination of the complex socio-technical systems associated with military operations.

### *Actor Network Theory*

Actor network theory is a theoretical perspective that has evolved to address the socio-technical domain and in particular the conceptualization of the 'social'. This perspective challenges the way we think of the human and non-human. The application of the ANT perspective (terms and concepts) has been instrumental in revealing insights within such fields as information technology, organizational theory, geography, medical anthropology and psychology. ANT treats both human and machine (non-human) elements in a symmetrical manner, thereby facilitating the examination of the situation (such as an accident) where Callon (1999:183) argues, 'it is difficult to separate humans and non-humans, and in which the actors have variable forms and competencies'. As noted by Ashmore, Woolfitt and Harding (1994:735), through ANT 'the assumption of the ontological primacy of humans in social research and theory is suspended. Non-human entities, traditionally overlooked in sociological accounts of the social world, take their rightful place as fully fledged actants in associations, relations, and networks'.

Fundamental concepts within ANT are the conceptualization of the *actor* and the *network*. An actor network as described by Latour (1987), Callon (1986a, 1991) is characterized as a network that is inherently heterogeneous, where the relations between the actors are important, rather than their essential or inherent features. The actor, whether technical or non-technical, is examined within the context of a heterogeneous network. In fact, the actor is a network in itself '… in the same way, elements in a network are not defined only by their "internal" aspects, but rather by their relationships to other elements, i.e., as a network' (Aanestad and Hanseth, 2000:360). The actors or actants of ANT can be humans, organizations, cultures, ideas, or inanimate objects and are described in terms of their interconnected network of relations. For example, Law (1987:11 4–116) describes the Portuguese carrack as simultaneously an actor within a much wider network 'such as the spice trade, and a network of wood planks, mast(s), sailcloth, crewmembers, investors, wind, stars, and navigational equipment, etc. In turn, each of these actors may be regarded as networks, and so on'.

As a piece of research informed by sociology, complex systems are seen as heterogeneous actor networks that consist of a particular configuration of more or less aligned human and non-human components. Within this conceptualization, actors may have different interests and agendas that are inscribed in both material and social actors/arrangements such that they enrol other actors through the process of translation (Rolland and Aaenstad, 2003). Examination of actors such as those characterized traditionally as technologies, facilitates an exploration of how these 'actors' mediate action and how they are entangled in local techno-social configurations. With regards to the context of a cockpit filled with computers, glass displays and the pilot, we leverage the comments from Harbers (2005:10) who asks the question, 'Where does one draw the line between man and machine, between human responsibility and technical inevitability?' Harbers (2005:10) argues that 'we are confronted here with a hybrid situation in which human beings and technology are tightly interwoven – a mixture, a muddle of man and machine'. We address these questions and arguments through the concept of the 'hybrid collectif' that emerges from the analysis. The three elements from ANT that are particularly relevant to the study of trust and teamwork in military operations include the principal of symmetry; the focus on actor networks and dissolving dualisms; and the emphasis on processes of translation (van der Duim, 2005:86).

Fundamental processes within ANT are inscription and translation. Inscription refers to the way technical artefacts embody patterns of use: Technical objects thus simultaneously embody and measure a set of relations between heterogeneous elements (Akrich, 1992:205). Monteiro (2000:77) argues that although 'inscription' might sound deterministic, 'the artifact is always interpreted and appropriated flexibly, the notion of inscription may be used to describe how concrete anticipations and restrictions of future patterns of use are involved in the development and use of a technology'. Inscriptions enable action at a distance by creating 'technical artifacts' that ensure the establishment of an actor's interests such that it can travel across space and time and thereby influence other work (Latour, 1987). Inscripted artefacts such as texts and images are central to knowledge work (Wickramasinghe et al., 2007:270) and thereby can shape sensemaking, decision making, action and are particularly relevant with distributed situational awareness, which places emphasis on the interaction between people and the artefacts they use (Stanton, 2010:2).

The process of translation has been described as pivotal in any analysis of how different elements in an actor network interact (Somerville, 1997). As a transformative process, translation emphasizes 'the continuous displacements, alignments and transformations occurring in the actor network' (Viseu, 2005:115). Translation rests on the idea that actors within a network will try to enrol (manipulate or force) the other actors into positions that suit their purposes. When an actor's strategy is successful and it has organized other actors for its own benefit, it can be said to have translated them. Translation as argued by Callon (1991:143) is 'embodied in texts, machines, bodily skills [which] become their support, their more or less faithful executive'. Hernes (2005:117) argues that translation can be regarded as 'negotiations, intrigues, calculations, acts of persuasion and violence, thanks to which an actor or force takes, or causes to be conferred on itself, authority to speak or act on behalf of another actor or force'. The actor network is therefore comprised of human and non-human elements through a series of negotiations such that actors seek to impose definitions of the situation on others (Callon, 1986b).

*Trust*

Exploring the notion of trust within the actor network sheds light on understanding situation awareness and decision-making processes that are examined within the case studies. The investigation of trust as a phenomenon crosses many domains of inquiry such as economics, political sciences, personality research and social psychology. As cited in de Vries (2005:5), each of these domains treats the topic in a contextual manner viewing trust in various schema '... whether it is seen as a dependent, independent, or interaction variable, whether it is static or dynamic, or whether it is studied on the institutional, group or individual level'. As described in Lee and See (2004:52), recognition of the importance of trust as a subject of inquiry has grown over the last number of years in recognition of its importance in shaping decision making, cooperation and communication. In organizational theory trust has emerged as a central topic, as noted in contributions from Kramer and Tyler (1996), Jones and George (1998), and Corritore, Kracher and Wiedenbeck (2003). In terms of understanding the role of trust in mediating human–automation interaction, some researchers have focused on trust as an attitude or expectation defining trust as: 'expectancy held of an individual that the word, promise or written communication of another can be relied upon' (Rotter, 1967:651); 'expectation related to subjective probability an individual assigns to the occurrence of some set of future events' (Rempel, Holmes and Zanna, 1985:96); 'expectation of technically competent role performance' (Barber, 1983:14). As reported in Riegelsberger, Sasse and McCarthy (2005:387), 'trust in technology is of particular importance for delegating to or relying on decision aids or software agents (Muir, 1987; Milewski and Lewis, 1997; Dzindolet et al., 2003)'. Trust in technology is characterised by Lippert and Swiercz (2005:341) as 'an individual's willingness to be vulnerable to a technology based on person-specific expectations of the technology's predictability, reliability, and utility as moderated by the individual's predisposition to trust the technology'.

As noted in Lippert and Swiercz (2005:342), the notion of trusting an inanimate object is not new. Giffin (1967) suggests that trust can be bestowed on a person, place, event, or object. In another effort along these lines, Muir (1987, 1994), and Muir and Moray (1996) employed an interpersonal approach to better understand the nature of trust between humans and machines and to determine the factors affecting this one-sided trust relationship. Lippert and Swiercz (2005:343) present the argument that 'trust as an expectation is tied to the notion that a technology will function in a consistent manner at a future time leading to an individual's assessment that the technology is predictable'. It has been argued that trust mediates not only relationships between people but also between people and automation and has been shown to affect reliance (Lee and See, 2004:51). The concept of system trust that is explored in this work 'can be seen as a special case of interpersonal trust' (de Vries, 2005:5) and is developed within the actor network perspective recognizing the inherent symmetry of the actor network, privileging neither human nor non-human actors.

## Methodology

Actor network theory stems from ethnographic and case study methodologies and as such does not have specific methods associated with it, although Callon (1986a and b, 1991) and Latour (1996) do provide the guidance 'to follow the actor'. As argued by Law and Callon (1988:285), 'we are concerned to map the way in which they [the actors] define and distribute roles, and mobilize or invent others to play these roles'. By following the actors we begin to reveal the fundamental importance of 'objects' within the context of teamwork and trust and hence learn from the actors 'without imposing on them an a priori definition of their world-building capacities' (Dolwick, 2009:38). Within the framework of 'follow the actor', thematic analysis of the data was conducted. As described by Boyatzis (1998:6), 'thematic analysis is regularly used by scholars and researchers in literature, psychology, sociology, cultural anthropology, history, art, political science, economics, mathematics, chemistry, physics, biology, astronomy and many other fields'. Thematic analysis (Boyatzis, 1998) is a search for themes that emerge as being important to the description of the phenomenon (Daly, Kellehear and Gliksman, 1997). The challenge associated with opening the 'black box' is, as argued by Williams-Jones and Graham (2003:272), 'to "unpack" and better understand the underlying processes and components of actors and networks that may not be readily apparent'. To start following actor interactions, it is necessary to develop a preliminary sketch of the network (Figure 12.1).

Combining the central tenets of ANT, the researcher follows the actors analysing the inscription and translation processes, thereby tracing and revealing the inherent interconnectivity and complexity resident within the network construct. Mind mapping (Buzan and Buzan, 2006) proved to be a valuable tool to facilitate the visualization of the interrelationships and connections within the network space, thereby giving rise to the emergence of insights from the coding and thematic analysis processes. These insights were further analysed to support the argumentation and explored using Anticipatory Failure Determination (AFD).

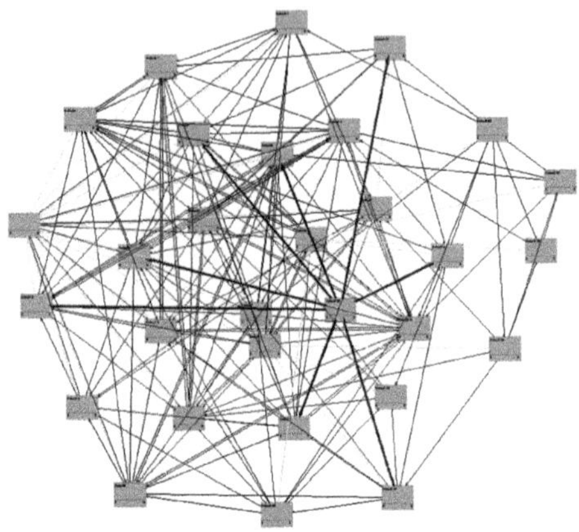

**Figure 12.1 Actor network visualization**

A spiral development approach facilitated the ANT methodology in that it allowed a process whereby increments of collecting, coding and analysis facilitated a conceptual convergence regarding pilot error and fratricide. The phases of thematic analysis: data familiarization; generating initial codes; searching for themes; defining and naming themes; and producing a report were integrated within the spiral development approach in Figure 12.2.

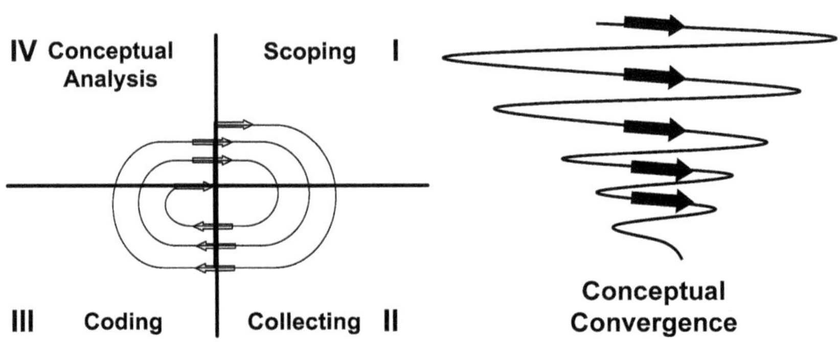


**Figure 12.2 Top view of spiral research framework (right) and side view of spiral research framework illustrating conceptual covergence**

As suggested by Latour (2005:9 7–120), the ANT research entails five sources of uncertainty, connecting the nature of: groups, actions, objects, facts and how to write research accounts.

1. The nature of *groups* concerns identifying what actors were assembled together. This was accomplished through examination of the sources of evidence. In each of the case studies, the relations between the actors were mapped and coded for common themes that were identified in the data (written reports, video and audio).
2. The nature of *actions* concerned the examination of agency within the actor network. This was accomplished by tracing the actor network and analysing the inscription and translation processes in this relational rhizome.
3. The nature of *objects* concerned recognizing the participation of non-humans in the course of action associated with fratricide. We examined which objects were being enrolled, mobilized or dispatched? Were these objects making a difference in the course of action?
4. The nature of *facts* concerned, within the context of fratricide, '...which facts were being disputed and made matters of concern? Which ones were being challenged, and which ones were standing up to and surviving those challenges?'
5. What characterizes an ANT analysis is the way of accounting for the social. As Latour (2005:131) argues here, social is not a special ingredient, or domain of reality, implying humans only. Instead, it refers to associations of radically heterogeneous actors.

Due to the arising complexity of the network analysis, it was necessary to identify key moments within the context of the case study and trace the relations from these points. These were derived from the linear event based descriptions of the accident aetiology contained in the case study material drawing upon the lessons learned from Tatnall (2000) and applications of ANT such as Callon (1986a, 1986b), Law (1987), Law and Callon (1988), Latour (1993), the work contained in Aanestad and Hanseth (2000), Aanestad (2003), and Czarniawska and Hernes (2005). The relational analysis was informed through supporting material such as Why Because Analysis (WBA) (Ladkin and Stuphorn, 2004), Systemic Theoretic Accident Mapping Process (STAMP) (Leveson, 2002; Leveson, Allen and Story, 2002) and Snook (2000).

As noted by Garson (2009:267), 'After years at the periphery of the social sciences, simulation is now emerging as an important and widely used tool for understanding social phenomena. Through simulation, researchers can identify causal effects, specify critical parameter estimates, and clarify the state of the art with respect to what is understood about how processes evolve over time'. Simulation facilitates exploration of assumptions that shape our models and understanding of complex systems. To inform the concepts of sensemaking and situation awareness and to facilitate understanding regarding the ANT processes of inscription and translation, insights were drawn from distributed simulations JSMARTS II and the MALO Project.

JSMARTS II was a limited scope experiment conducted to examine the JSMARTS principle of a Modelling and Simulation 'pick-up game' within the context of a dirty bomb in Ottawa. The main purpose for JSMARTS II was to capture how existing defence M&S capability can be used in civilian emergency management environments, yielding new capability in emergency management simulation and analysis that will be of interest to both communities. The JSMARTS II experiment provided an opportunity to examine and evaluate the dimensions of Situation Awareness and explore the processes of inscription and translation within a controlled synthetic environment. The distributed simulation, utilizing a High Level Architecture (HLA) facilitated the problem space for the scenario. The MALO Project developed and demonstrated a limited synthetic environment (SE) of the Maritime Air Littoral Operational Environment required for a Maritime air platform to operate as part of a task force. Two specific applications shaped the MALO SE: namely multi-platform, multi-sensor, Anti-Submarine Warfare (ASW) and Anti-Surface Warfare Task Group support operations; and the application of airborne sensors to coastal and overland surveillance and targeting. The Anticipatory Failure Determination (AFD) modelling environment, described in detail in Masys (2010) facilitated the relational mapping of the space of possibilities combining both the results from the thematic analysis and following the actors. The relational mapping described resonates with the propositional networks described in detail in Stanton et al. (2009).

The case study analysis was conducted on a number of fratricide incidents shown in Tables 12.1 and 12.2 and described in detail in Masys (2010).

**Table 12.1 Primary fratricide case studies**

| Date | Incident | Nations involved |
|---|---|---|
| 17 Feb 1991 | Apache helicopter/Bradley fighting vehicle | US/US |
| 17 Apr 2002 | F-16/Canadian soldiers | US/Canada |
| 14 Apr 1994 | F-15/Black Hawk helicopter | US/US (multinational) |

**Table 12.2 Secondary fratricide case studies**

| Date | Incident | Nations involved |
|---|---|---|
| 5 Dec 2001 | B-52/US soldiers | US/US (multinational) |
| 22 Mar 2003 | Patriot missile battery/RAF Tornado | US/UK |
| 28 Mar 2003 | A-10/armoured vehicles | US/UK |

The US Patriot missile fratricide incident of 22 March 2003 is of particular interest as it clearly reflects team and trust issues within a socio-technical domain. A Tornado had been operating as part of the RAF Combat Air Wing based at Ali Al Salem in Kuwait. All flight preparations including briefings, start up, take off and the operational phase of the sortie were all completed as planned. As part of the preflight checks, the Identification Friend or Foe (IFF) system was checked by the ground crew and confirmed to be working correctly. The Patriot battery crew were monitoring for Iraqi Tactical Ballistic Missiles when the Tornado was tracked by their system. The Patriot system indicated that an Anti-Radiation Missile was coming directly towards them. The track (Tornado) was interrogated for IFF but there was no response. Having met all classification criteria, the Patriot crew launched the missile, and the Tornado, mistaken for an Anti-Radiation Missile, was engaged in self-defence. The Patriot crew had complied with extant self-defence Rules of Engagement for dealing with Anti-Radiation Missiles (MOD, 2004:2). It is clear that the immediate cause of the accident was that a Patriot missile destroyed the Tornado. Contributory factors associated with the Patriot fratricide included: Patriot Anti-Radiation Missile classification criteria; Patriot Anti-Radiation Missile Rules of Engagement; Patriot firing doctrine and crew training; Autonomous Patriot battery operation; Patriot IFF procedures; Tornado IFF serviceability; aircraft routing and airspace control measures, and Orders and Instructions.

The development and fielding of the Patriot air defence missile system in the early 1980s represented a significant improvement in operational capability that allowed the US to wage a computer-aided air battle by incorporating decision-making logic into the weapon system itself – as opposed to a separate C2 system. This capability, through its decentralized engagement logic, permitted operators to handle a larger number of threats and speeded engagements by automating portions of the decision-making process. The system received accolades resulting from its operational success during Operation Desert Storm and most recently during Operation Iraqi Freedom (OIF). In both Gulf wars, missiles were successfully engaged by Patriot employed in a fully automatic, operator-monitored mode. However, these successes were marred by an unacceptable number of fratricidal engagements attributable to track misclassification problems, particularly during OIF (Hawley, Mares and Giamannco, 2005:2). The Combat Identification (CID) process inherent within the Patriot system identifies hostile missiles through their flight profile and other characteristics, including the lack of an IFF response. The results of the investigation noted that the criteria used to identify hostile missiles should have been better managed, based on the known threat from Iraq, and concluded that the generic Anti-Radiation Missile classification criteria programmed into the Patriot computer were a contributory factor in the accident. The Board concluded that the ROE associated with the Patriot system was not robust enough to prevent a friendly aircraft being classified as an Anti-Radiation Missile and then engaged in self-defence.

As described in Hawley et al. (2005), Patriot crews are trained to react quickly, engage early and to trust the Patriot system. In hindsight, had the crew delayed firing, the Tornado would probably have been reclassified as its flight path changed. The crew had about one minute to decide whether to engage. The crew was fully trained, but their training had focused on recognising generic threats rather than on those that were specific to Iraq or on identifying false alarms. The Board concluded that both Patriot firing doctrine and training were contributory factors in the accident. The Patriot crew was operating autonomously, with a primary role of protecting ground troops from missile attack, but the Rules of Engagement allowed the battery to fire in self-defence. A critical component of its capability lies with its communications suite which was apparently still in transit from the US, therefore contact with the Battalion HQ and other units was through a radio relay with a nearby battery, which was equipped with voice and data links to and from Battalion HQ. The lack of communications equipment meant that the Patriot crew did not have access to the widest possible 'picture' of the airspace around them to build situational awareness. The Board considered it likely that a better understanding of the wider operational picture would have helped the Patriot crew, who would then have been more likely to identify the Tornado as a friendly track, albeit one without a working IFF. The Board concluded that the autonomous operation of the Patriot battery was a contributory factor.

As described in Hawley et al. (2005), IFF is a system designed to identify automatically whether or not a particular asset, such as an aircraft, is a 'friend or foe'; civilian Air Traffic Control also use it to identify and track aircraft. The system works as a challenge and reply whereby a signal is sent from the ground or air to the aircraft, which then replies with the appropriate code, thereby providing identification. There are five different modes of IFF, which can work in parallel or alone. These include Mode I (an unencrypted code, which was used in Iraq by all the Coalition aircraft) and Mode IV (an encrypted form of IFF). Investigation showed that the Patriot battery's IFF interrogator for Mode IV was working throughout the engagement period, but that Mode I codes were not loaded. The Board believed that autonomous operations without voice and data connections to and from Battalion HQ might have contributed to the difficulty the battery had in receiving the Mode I IFF codes. The Board concluded that the lack of IFF Mode I codes increased the probability of the accident, and was therefore a contributory factor. The Board considered IFF serviceability, potential IFF failures and aircrew actions relating to the IFF. Following initial investigation, it became apparent that certain power failures associated with the IFF may not be displayed to the crew. The most likely explanation for the absence of an IFF response was that there had been a power supply failure. The Board recommended that further work be conducted to research the failure modes, reliability and serviceability of the Tornado IFF system.

As described in Hawley et al. (2005:25), 'the Boards of Inquiry examining the root causes of Patriot fratricide incidents during Operation Iraqi Freedom concluded that the training provided to van crews was a contributing factor'. The training practices of the air defence community were criticized for their emphasis on rote drills rather than high level judgment. 'What this means is that much pre-OIF Patriot training was reduced to a stimulus-response exercise with little intervening thought or judgment: If you see X … Then do Y. A rote, crew-drill approach to training might be appropriate for many aspects of air defense operations (e.g., march order, emplacement, system set-up, etc.), but it is not suitable for air battle operations or management. These require a focus on adaptive decision making within a complex and dynamic tactical setting' (Hawley et al., 2005:25).

A shortfall, according to the Patriot System Performance Report (2005:2), 'was the lack of significant situational awareness in the combined air defense system, which involved major systems such as Patriot, AWACS and AEGIS. It is assumed that data are routinely communicated from one system to the other, that targets are correlated, and target information is shared and assimilated by all'. The assumption is a long way from reality. In fact, 'the communication links, the ability to correlate target tracks by disparate sensors, and the overall information architecture are simply not there' (Patriot System Performance Report, 2005:2).

The RAF Board of Inquiry was carried out in parallel with a US investigation. The Board of Inquiry report concluded that the contributory factors were complex, many and various, and has made a series of recommendations which are currently being implemented (NAO, 2006). Major General Vane (President of the Board of Inquiry) was convinced that human performance issues were part of the problem associated with the fratricide incident. As noted in the Hawley and Mares (2007:1), 'he was particularly concerned by what he termed a "lack of vigilance" on the part of the Patriot operators along with an apparent "lack of cognizance" of what was being presented to them on situation displays with a resulting "absolute trust in automation".'

This case study brings to the forefront how technology is not an external part of the human system but is actually integrated into the SA and decision-making process. Issues of trust and expectations thread through this case from ROE through the software and hardware and training and doctrine. It reflects the entangled nature of the accident aetiology associated with complex socio-technical systems.

## Discussion

The ANT worldview presents a methodology that stems from the interpretation of the word 'social' as described in Latour (2005). In the broadest sense, 'social means association. This is after the Latin word socius, meaning a companion or associate, with the root, sequi, meaning "to follow"' (Dolwick, 2009:21). As such, the social is construed as a connection, and interaction which may include 'plants, animals and material artefacts as well as humans' (Dolwick, 2009:2 1–22). One of the critical concepts that shape the analysis through the ANT lens is that actors are regarded as relational effects. What is unique about the ANT lens is that it draws our attention to the system interaction over multiple time-spaces exhibiting a disproportionality of 'causes and effects' and hence informs ideas regarding teamwork and trust. As argued in van der Duim (2005:88), 'It does not make sense to ignore materials and to treat them separately, as though they were different in kind: the characterization of materials is just another relational effect.' The significance of trust in this work stems from the issue regarding factors shaping decision making. For example, technology trust is linked to socio-technical trust in that 'human trustors are known to treat technological artefacts in similar ways as they treat human ones' (Riegelsberger, 2005:71). From the ANT perspective, trust resonates with matters pertaining to inscription and translation and their effects throughout the actor network. In particular, we note the extension of trust beyond the person to person to include the inanimate, thereby reflecting the impact of the technical on the 'social'.

Through an iterative process of coding both within case and cross-case analysis, as well as input from the simulation exercises, emergent themes were captured via the process of thematic analysis in which the themes become refined through an iterative process, thereby resulting in overarching concepts. Figure 12.3 shows a high level view of the problem space derived from the cross-case and within case analysis, thematic analysis and AFD analysis. AFD, through a structured methodology rooted in the Theory of Inventive Problem Solving (TRIZ), facilitates an examination of the problem space through failure analysis and failure prediction and thereby provides a more explicit and detailed exploration of the actor network, relations and processes of translation and inscription. Four very general types of actor emerge from the analysis: human beings, with the skills and knowledge that they generate and reproduce; artefacts, which include all the

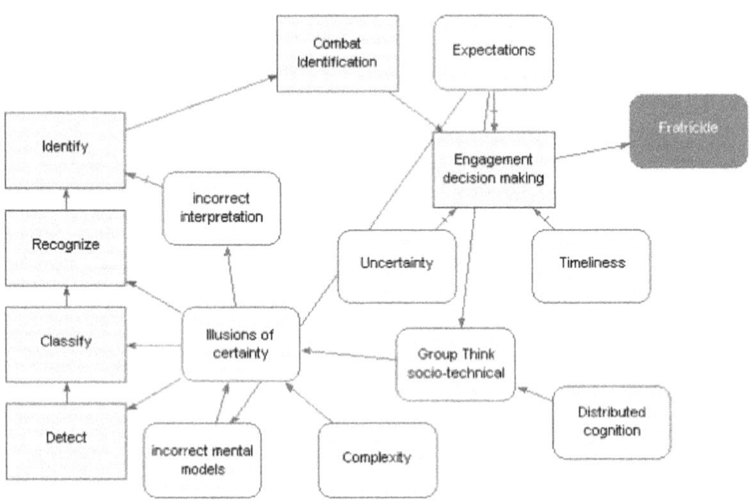


**Figure 12.3 Thematic view of problem space**

non-human entities that facilitate performance of a task; texts and inscriptions, which include everything that is written or recorded (such as SOPs, ATOs, ROE) , as well as the channels through which they circulate (such as command and control processes); and institutional authority, which is embedded in regulations and power relationships (Gherardi and Nicolini, 2000:16).

In defining the actor network, technology, as an actor within the socio-technical system, is not regarded as an object in itself but rather as a relational entity. Michael (2003:130) argues that 'technologies work as they do because they are composed of complex heterogeneous distribution – of assemblages – of humans and non-humans'. Stemming from the case study analysis, the pilot emerges as 'an effect generated by a network of heterogeneous, interacting, materials' (Law, 1992:381). In defining the actor network, the accident aetiology therefore becomes this entangled network of heterogeneous elements, a *hybrid collectif* (Figure 12.4). Law (1992:381) argues that:

> … what counts as a person is an effect generated by a network of heterogeneous, interacting, materials … But converted into a claim about humans it says that people are who they are because they are a patterned network of heterogeneous materials. If you took away my computer, my colleagues, my office, my books, my desk, my telephone I wouldn't be a sociologist writing papers, delivering lectures, and producing 'knowledge'. I'd be something quite other – and the same is true for all of us.

Similar to the thought experiment described in Callon and Law (1997:171) describing 'Andrew the strategist', from the relational tracing we selectively begin removing actors within the actor network, thereby conducting a sensitivity analysis of the accident aetiology by exploring the space of possibilities. The pilot ceases to exist with the removal of the sensor systems, navigation systems, communication systems, the elements necessary to complete the Combat Identification process. What we discover is the pilot and essentially 'pilot error' is a heterogeneous network: 'person' + aircraft + supporting crew and technicians + orders, SOPs, ROE + avionics + sensors + virtual team members + communications + training + doctrine + air force culture + work of engineers + legal council + politicians. What this reveals is that action, derived from cross-case and within case analysis, takes place in a 'hybrid collectif' that is comprised of entangled human actors as well as non-human actors in multiple ways. Viewed from this perspective, tools (such as the hardware and software) that are embedded in the actor network are, as Callon and Caliskan (2005:18) remark, not just things that are used to achieve certain ends: 'They contribute to the making of the universe of possibilities that make action itself.' It is this very relational network of actors that creates the possibility for action and redefines the notion of team to include the non-human and that which is temporally and spatially displaced.

**Figure 12.4 Hybrid collectif characterising the socio-technical system**

To facilitate exploring the actor network, 'synthetic environments' were employed whereby the virtual world is comprised of a distributed network architecture. These distributed simulations represent an instantiation of the actor network. The HLA paradigm of the distributed simulations represents a suitable analogy for the actor network. Like the actor network, the federates are relationally interrelated within the simulation. The federates have inherent inscriptions that shape the interactions within the distributed simulation and result in an emergent behaviour. What emerges from a study of these simulations is how federates, representing black boxes in a network, establish expectations based on the apparent simulation fidelity that shape sensemaking, action and decision making. For example, in the JSMARTS II, the radiation dispersion model was inscribed with capabilities and limitations that shaped the course of the simulation exercise. The fidelity associated with the 3-D models (Figure 12.5), translated and established expectations regarding the fidelity of the radiation models, thereby establishing a belief that the radiation models take into consideration the complex material and structural nature of the city buildings in terms of radiation attenuation. Similarly, in the MALO simulation, the federates come 'hardwired' with their inherent limitations, capabilities and biases. This very choice of a federate capability actually sets inherent parameters within the simulation that affects the tactical employment of underwater sound in the prosecution of a terrorist subsurface threat. The operator using the simulation thereby approaches the exercise with expectations that have been inscribed into him through a process of translation (training, both experiential and text based). These inscription and translation processes emerge from the Federation Object Model (FOM) and Simulation Object Model (SOM) that characterizes HLA distributed simulations.

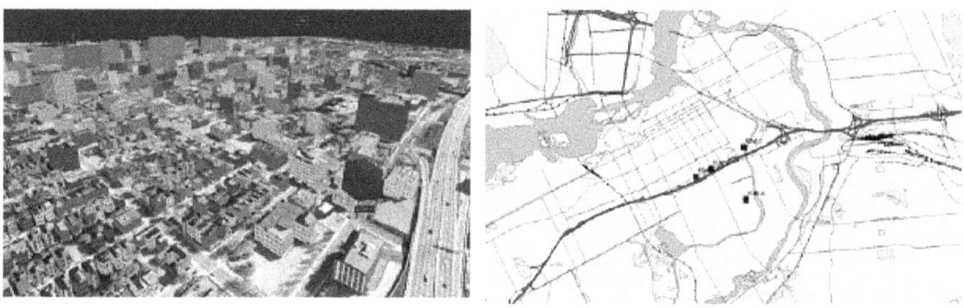

**Figure 12.5a Simulation views of problem space**

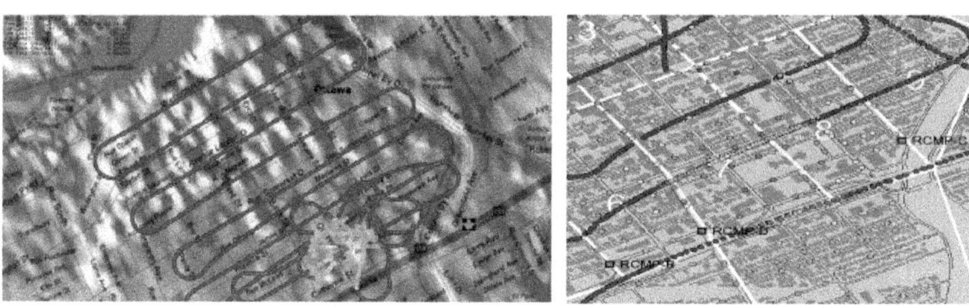

**Figure 12.5b Images of synthetic environment from JSMARTS II**

From participant observation, what emerges from this exercise is the realization that technology mediates and shapes the human experience and relations to their world. Taking these observations from the M&S world and applying them to the fratricide case studies reveals how actors such as the ROEs, Air Tasking Orders and supporting physical actors (GPS, IFF) all shape the actions that emerge from the actor network. Mediation can therefore be considered the mixing of humans and non-humans in this actor network. It is through this mediation (mixing) that voices become deleted, the visible become hidden, a black box is created. The process of black boxing (intentionally or unintentionally) makes the actors and their relations opaque (Latour, 1994a:36) as represented in the simulations whereby the federation (consisting of federates) represents a black box.

## Hybrid Collectif

The ANT relational analysis and distributed simulations highlights that action takes place in 'hybrid collectifs' (Callon and Law, 1995) that entangle human actors as well as non-human actants in multiple ways. Drawing upon the definition of 'social' in terms of an association, the concept of team emerges as a relational hybrid of human and non-human (technological, natural, material) elements simultaneously. Law (1994:23) argues that the social world is 'materially heterogeneous', 'there would be no social ordering if the materials which generate these were not heterogeneous … Left to their own devices human actions and words do not spread very far at all'. The tenet of 'free association' within ANT, rejects a priori distinctions between 'the social' and the non-social, and thereby facilitates an examination of the ways in which people and things are associated in networks. From this viewpoint we see that 'there are not a few hybrids but that there are only hybrids' (Crawford, 1993:261).

As entities are relationally defined, so action arises from the distributed set of competencies resident within the actor network in which humans and non-humans are both full participants. This sets the stage for understanding trust and teamwork in the actor network context. The pilot is part of a web of relations linking heterogeneous elements (of human and non-human entities). The notion of teamwork associated with the fratricide case studies become associations stitched together whereby the entities become enrolled, combined and disciplined within networks, thereby gaining shape and function (van der Duim, 2005:92).

Translation and mediation comprise what Latour (1999a:186) refers to as 'programs of action'. Latour (1994a:35) argues that 'action is simply not a property of humans but an association of actants'. This is a key point. Trust is viewed not from a human-centric perspective but rather recognizes that trust is derived from a network of actors and relations. In the case of the Patriot system incident, translation processes can be seen to emerge from the analysis associated with the automation and Human Computer Interface (HCI) that shapes action and inaction. It is associated with the 'presumed' accuracy associated with the navigational systems, the authority and assumed capabilities associated with the command and control infrastructure. The case studies reflect how actors such as these have an inherent accuracy, certainty and authority that translates (forces) the actor (as part of the hybrid collectif) to act in a certain manner. This is a critical concept in the analysis of teamwork and trust. With consideration of the accident aetiology associated with fratricide, rather than the question, 'How could they not have known?' with its hindsight bias, we approach the problem space with the insightful question what actor network dynamics, translation and inscription processes precipitated and contributed to the accident in the case studies ... What does this say about the nature of teamwork and trust?

It is recognized that the actors are not passive entities but participate in the creation of possibilities. To paraphrase Baygeldi and Smithson (2004:118), technology can be used as an instrument of influence and hence embody micro politics of power. The actors, through the distributed nature of agency within the network, 'vary in the extent to which they influence or resist the influence of other entities' (Somerville, 1999:10). What is important to emphasize as reported by Somerville (1999:10):

> Not only are humans and non-humans to be seen within the same (conceptual and terminological) framework, but micro-actors (individuals, computers, etc) and macro-actors (institutions, corporations, governmental organizations, etc) are to be seen in this way as well.

This is particularly evident in the simulation experiment of JSMARTS and MALO. It reveals:

> Technologies are not simply passive and are never value neutral, but always exist in value-laden social and technical relations. During the design phase, objects have embedded within them a 'script' or set of instructions that determine how the technology will function and the extent to which it may be shaped by other actors (Williams-Jones and Graham, 2003:276).

Within these synthetic environments, the very nature of the High Level Architecture and the distributed nature of the models (each scripted for a specific purpose: but not necessarily designed for this specific application) parallels the actor network world view. The presentation of the information from the simulations (as actors) shapes the perception and challenges the mental models of the observer (also an actor) that will shape SA and decision making. As noted in Faraj, Kwon and Watts (2004:194), supporting this notion of inscription '...beliefs can arise either from the technology histories of particular actors or from interdependent relationships among multiple actors'. This is particularly demonstrated in the simulation experiments. The belief regarding the validity and fidelity of the models (federates) is inscribed into its development histories and is evidenced through another actor (the corresponding documentation that details the model's capabilities and limitations). Belief therefore becomes interrelated to the pedigree of the model, its verification and validation (V&V) evidence (Masys et al., 2008) and its representation. Arising from the simulation experiment and case study analysis, it becomes apparent that 'objects and other non-human entities do affect human behaviour' (Williams-Jones and Graham, 2003:273). Artefacts can be designed to 'replace human action and constrain and shape actions of other humans' (Latour, 1992:225).

Through the ANT lens, SA derives from a distributed collaborative environment, thereby emphasizing the importance of the larger picture, including the material, social and organizational dimensions. The federate representations within the simulations shape the way the operators viewed the world. As a distributed simulation, the federates (HLA) come hardwired with their inherent characteristics. Their collective behaviour arises from the integration of all the elements (complete with built-in assumptions) inscribed. The effectiveness of the simulation is dependent on the accuracy associated with its interrelationships as compared to the real world. In post exercise debrief it was revealed that certain inherent limitations and constraints existed within the federate models by virtue of the conscious and intentional development of the HLA federation (in other words the characteristics of the federate models were inscribed with specific qualities and fidelity in support of the developer's agenda). With a superimposed 3D model of the environment (Figure 12.5), an assumption (illusion of certainty) was made as to the reliability, correctness and fidelity of the actor network. Within the domain of M&S, the illusions of certainty are best captured by the adage 'Garbage in, Hollywood out' (Roman, 2005:1). It represents how inscribed dysfunctionality within models can be 'represented' as having greater fidelity and certainty than it actually has. It represents an emergent behaviour that is realized within an actor network that includes both humans and non-humans. The high precision of electronically displayed information can sometimes obscure the actual uncertainty or ambiguous nature of the underlying data and thereby affect SA. The distributed simulation provides a window into the complexity in the real world in terms of the interdependencies and interrelationships. In particular the temporal displacement associated with the model development from the actual execution emphasizes the inscription and translation processes as they transcend the proximate. In terms of SA, the experiment revealed how the elements of the simulation, including hardware, software, sensors, programmers and participants, together shaped the situation awareness through its representational fidelity. The AFD analysis reveals how SA within the SE is distributed among the network of heterogeneous elements and how small changes or 'discrepancies' within the network can propagate and affect the SA (Figure 12.6). The pathways to an accident are complex, interrelated and interdependent; relations thereby necessitate cohesion and alignment of the actors.

As argued by Verbeek (2005:130), the concept of delegation indicates that programmes of action can be 'inscribed' into artefacts. Within the context of the case studies, technical mediation expands our notions of action and experience. Actors within the network, such as a technical artefact, co-shapes the human world relations by giving shape not only to people's actions but also to people's experiences. The experience of the pilots and patriot crew was mediated by their participation within this hybrid of human and non-human actors that relationally transcend linear temporal and spatial conceptualizations.

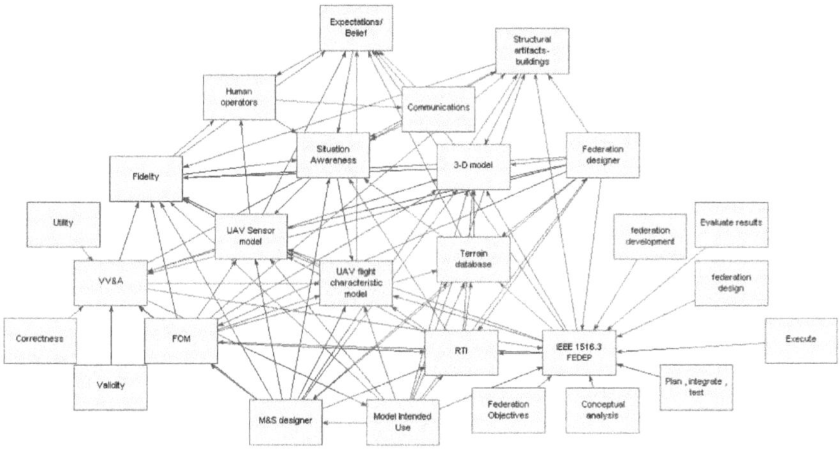


**Figure 12.6 SA relational mapping within JSMARTS**

Tracing from ROE to the fratricide incident, actors were aligned to enable the political, military and legal agendas reflected in the ROE (an act of translation and inscription) shown in Figure 12.7. This actor network enabled action at a distance (Latour, 1992:225). The physical actors within the hybrid collectif determine certain actions through their entanglement with the informational domain (ROE, SPINS, ATO) and the human. Without this alignment, the ROE would not be able to be enabled. For example the employment of IFF (representing the ability to identify and classify an object) can be viewed as ‘a substitute for the action of people and is a delegate that permanently occupies that position of a human’ (Latour, 1992:234). In the Patriot case study, the IFF represented this ‘identification’ supporting the application of the ROE. The GPS, with its inherent accuracy and reliability, is a substitute for the map-reading duties of the aircrew. The GPS is delegated this position in the actor network.

Trust emerges in terms of expectations, knowledge and assumptions about how other actors are likely to behave in the future. In the case of the Patriot fratricide, trust can be viewed to simplify and reduce the complex set of expectations that we use to predict how the other actors will behave, which will in turn guide and shape sensemaking, decisions and actions. As the case studies show, issues of trust come into play in situations that contain risk, vulnerability, uncertainty and interdependence. With the digitization of the battlespace, such actors as expert systems and automated decision aids gives rise to questions regarding trust. What emerged from the case study analysis are the illusions of certainty that permeated the actor network space, that like a socio-technical groupthink (Masys, 2010) supported the decision and subsequent action that precipitated the fratricide. Through the ANT lens, the definition of the team (actor) is expanded to include the technical artefacts that make action possible. These artefacts are inscribed with

**Figure 12.7 High level ROE actor network**

certain behaviours and characteristics that shape action, since action cannot be considered without their participation.

Illusions of certainty have everything to do with expectations. As Weick and Sutcliffe (2007:23) argue within the context of organizations:

> … expectations are built into organizational roles, routines, and strategies. These expectations create the orderliness and predictability…. Expectations, however, are a mixed blessing because they create blind spots. Blind spots sometimes take the form of belated recognition of unexpected, threatening events. And frequently blind spots get larger simply because we do a biased search for evidence that confirms the accuracy of our original expectations.

In a socio-technical space, when individuals equate general expertise with situational knowledge they create and rely on unrealistic expectations of those 'experts' (Barton and Sutcliffe, 2009:1341), be they human or non-human, present or absent. This 'expertise' denoted as accuracy and reliability participates in the process of inscription and translation whereby actors are inscribed with an apparent expertise, thereby persuading (forcing) other actors to recognize this attribute and defer to them. This crosses all domains of human, physical and informational and is reflected in the ROE, SOPs, hierarchical structure of air operations with respect to command and control and technical hegemony that shapes perception and decision making. The Patriot fratricide is an example of this expectation that has been delegated to the system by virtue of its inherent 'believed' reliability and accuracy rooted in its 'authoritative' performance specifications. It is further inscribed in the SOPs and training, thereby translating behaviour to ensure processes are followed. These expectations are relationally rooted to the ROE that are inscribed into the system. Similarly, the Patriot case study highlights how unacknowledged system fallibilities and fascination with blind faith of technologies were implicated in the accident aetiology. As described in Hawley, Mares and Marcon (2010), system fallibilities were known in the 1980s; however, they were not satisfactorily addressed during system software upgrades, 'nor did information about them find its way into operator training, battle command doctrine, operating procedures, or unit standard operating procedures' (Hawley et al., 2010:306). Following the actors associated with this fratricide highlights how ROE, SOP and system reliability were dysfuctionally aligned, or rather were aligned through an illusion of certainty, thereby precipitating the fratricide incident. This illusion of certainty permeated the organizational culture which emphasized 'reacting quickly, engaging early, and trusting the system without question' (Hawley et al., 2010:306). As well, this trust in automation shaped the organizational management of personnel 'which tended to place inexperienced personnel in key battle command crew positions' (Hawley et al., 2010:306).

Mismatches between trust in the system and actual system performance result in inappropriate human monitoring and information-sampling behaviour resulting from translation processes associated with ROEs that permeate the network space and emerge within the SOPs. Such is the case with the Patriot fratricide in which the Patriot operator uncritically counted on the reliability of automated systems. This phenomenon is called complacency. Due to the opacity of automated systems, trust mismatches may occur which undermine the benefits of automation (as within the Patriot fratricide case in which overreliance on system capability precipitated the accident). Trust thereby emerges as a central component of safety performance.

Safety specific trust and distrust is modelled by Conchie and Donald (2008), representing how these two factors affect safety operations. Drawing upon this model, it is clear that the mapping of the trust and expectations as described within the context of the case studies highlights how dysfunctional trust and distrust affected safety performance. Trust emerges from the Patriot case in which personal responsibility was delegated to the automation of the system. Evidence from the case study supports the notion that both trust and distrust have a role in safety performance in operations. Information trust refers to a user's willingness to accept a given piece of information into a decision-making process when the use of 'bad' information could be a critical mistake. This is significant because as we begin to trace the actor network (follow the actors) what emerges is the process of translation that encourages other actors to the support dysfunctional trust. What becomes apparent is how, through the network analysis, the qualities of precision, accuracy and reliability are translated into operational capabilities and safety qualities. NVG, GPS, laser guided bombs, radar warning receivers, IFF, AWACS, training and SOP (a socio-technical team) all entangle their inherent and relational properties upon each other and the network as a whole to impose and stabilize certain characteristics, qualities and attributes of the actor network 'team'. Trust thereby becomes an inscribed feature of the actor network that shapes action.

Lee and See (2004:54) define trust as 'the attitude that an agent will help achieve an individual's goal in a situation characterized by uncertainty and vulnerability'. Trust in technology (or other actors such as ATO accuracy and timeliness) mediates the relationship between users' belief about the actor's capability and their reliance on the actor. Supporting the findings of Wang, Jamieson and Hollands (2008:292) and Jamieson, Wang and Neyedli (2008:27), it became apparent that trust in an actor (technology, ROE, SOP) acts as an attitude-shaping mechanism mediating the relationship between the users' belief about the acting capabilities and their reliance on the aid.

Black boxing of technology with inherent 'perceived' characteristics of certainty also black boxes the notion of trust within the actor network in terms of expectations that have been inscribed and translated. The Patriot incident reveal how the technological and organizational (process) solutions 'inscribe' a sense of reliability, set expectations and form a trust relation that permeates the actor network shaping standard operating procedures, orders and decision making. The translation processes within the actor network aligned actors, creating an illusion of certainty that within an actor network groupthink created a convergence that precipitated the accident. Mediating action is co-shaping what is happening. The mediation of action according to Latour has the form of 'prescriptions' that can be expressed in language as a 'script', a series of instructions on how to act (Verbeek, 2005:129). This is the basis of the argument whereby actors, a network of heterogeneous elements, co-shape what is happening. The pilot is no longer the centre of the accident aetiology. The socio-technical network becomes redefined in terms of a team defined by the network of heterogeneous elements.

## Conclusion

The application of actor network theory has been instrumental in opening the black box of pilot error. Actor network theory presents all entities (people, concepts and actions) as taking form and attributes as a function of their relation with other entities (Law, 1999). A network, as proposed in Baygeldi and Smithson (2004:119), 'can be described as a dynamic system of communication, cooperation and partnership between individuals and groups'. Within the context of this problem space and the actor network perspective, we expand upon this to remap the individual and group to a network of heterogeneous elements, both human and non-human, giving rise to a schema of interconnectivity and relationality. This work shows and supports the findings in the literature that: 'action is equivalent to specific and materially heterogeneous relations … hybrid collectifs. These relations, human and non-human, carry action, they exert it, and they modify it' (Callon and Law, 1997:172). What is argued is that if we wish to understand teamwork and trust then we need to explore the patterns of relations that reside in the socio-technical system.

We recognize from the analysis that:

- Non-humans have significance and are not simply resources or constraints.
  - Non-humans intervene actively to push action in unexpected directions: they emerge as part of a socio-technical distributed team.
- Entities are interactive effects.
  - They are networks of associations of human and non-human, thereby shaping trust and redefining the notion of team.
- Action results from the complex interactions resident within the actor network that is dynamically shaped by inscription and translation processes.
- The actor network lens reveals that action cannot be explained in a reductionist manner, as a firm consequence of any particular previous action (Callon and Law, 1997:172).

As noted in Brey (2005:76) the performance of technical artefacts and their inherent capabilities is not a product of the object itself but is shaped by the network of artefacts that are interconnected and interdependent. The hybrid collectif that emerges from the network reaffirms that artefacts that comprise the actor network cannot be understood in isolation. Thrift (1996:1468) makes a supporting observation:

> … no technology is ever found working in splendid isolation as though it is the central node in the social universe. It is linked – by the social purposes to which it is put – to humans and other technologies of different kinds. It is linked to a chain of different activities involving other technologies. And it is heavily contextualized.

Within the context of this work, the technology (actors) that is embedded within the network is relationally connected to the 'pilot' (actor), thereby defining a distributed team that shapes action and inaction. Technologies, texts, artefacts or non-human actors must be recognized as enablers of the network forming this hybrid collectif. The analysis reveals how the boundaries between humans and non-humans become ever more blurred. In this relational space, 'it makes no sense to talk of a machine, computer, technology in general than it does to talk of a "human" in general' (Graham, 1998:178).

Verbeek (2005:131) in his analysis remarks that 'Artifacts influence the way in which people do things, and this influence could be deliberately inscribed into them'. This is supported by the actor network analysis. What became apparent is a process of technical mediation, whereby the artefacts co-shaped the relational world (network space) by influencing or 'giving shape not only to people's actions but also to people's experiences' (Verbeek, 2005:139). Viewing pilot error as a socio-technical phenomenon (as a network of heterogeneous elements relationally interconnected) redefines the notions of team and trust. Law (1992:4) argues (but contextualized for this work) that pilot errors are never located in bodies and bodies alone, but rather that an actor is a patterned network of heterogeneous relations, or an effect produced by such a network. Law (1992:4) argues that 'a machine is also a heterogeneous network – a set of roles played by technical materials but also by such human components as operators, users and repair-persons. So, too, is a text. All of these are networks which participate in the social'.

The socio-technical domain is revealed as a complex, interconnected, relational, entangled state that emerges as a hybrid collectif of human and non-human actors. The ANT perspective reveals (in all case studies presented), that the notion of teamwork and trust emerges from the network of heterogeneous elements (without differentiating between human and non-humans).

## References

Aanestad, M. (2003) The camera as an actor design-in-use of telemedicine infrastructure in surgery, *Computer Supported Cooperative Work (CSCW)*, 12 (1), 1–20.

Aanestad, M. and Hanseth, O. (2000) Implementing open network technologies in complex work practices: A case from telemedicine. In: Proceedings from the IFIP WG 8.2 Conference IS2000, 10–12 June. Aalborg, Denmark: *Organizational and Social Perspectives on Information Technology*. (pp. 355–369), Kluwer Academic Publishers.

Akrich, M. (1992) The de-scription of technical objects. In: W. Bijker and J. Law (eds), *Shaping Technology, Building Society: Studies in Sociotechnical Change* (pp. 205–224), MIT Press: Cambridge, MA.

Ashmore, R., Woolfitt, R. and Harding, S. (1994) Humans and others, agents and things. *American Behavioral Scientist*, 37 (6), 733–740.

Barber, B. (1983) *The Logic and Limits of Trust*. Rutgers University Press: New Brunswick, NJ.

Barton, M.A. and Sutcliffe, K.M. (2009) Overcoming dysfunctional momentum: Organizational safety as a social achievement. *Human Relations*, 62 (9), 1327–1356.

Baygeldi, M. and Smithson, S. (2004) Ability of the actor network theory (ANT) to model and interpret an electronic market. In: J. Gupta (ed.), *Creating Knowledge Based Organizations* (pp. 109–126), Idea Group Publishing: Hershey, PA.

Boyatzis, R.E. (1998) *Transforming Qualitative Information: Thematic Analysis and Code Development*. Sage Publications: Thousand Oaks, CA.

Brey, P. (2005) Artifacts as social agents. In: H. Harbers (ed.), *Inside the Politics of Technology: Agency and Normativity in the Co-production of technology and Society* (pp. 61–84), Amsterdam University Press: Amsterdam.

Buzan, T. and Buzan, B. (2006) *The Mind Map Book*. BBC active, Pearson Education Group: Essex.

Calhoun, M.A. and Starbuck, W.H. (2003) Barriers to creating knowledge. In: M. Easterby-Smith and M.A. Lyles (eds), *Handbook of Organizational Learning and Knowledge Management* (pp. 473–492), Blackwell Publishing: Oxford.

Callon, M. (1986a) The sociology of an actor-network: The case of the electric vehicle. In: M. Callon, J. Law and A. Rip (eds), *Mapping the Dynamics of Science and Technology* (pp. 19–34), MacMillan Press: London.

Callon, M. (1986b) Some elements of a sociology of translation: Domestication of the scallops and the fishermen of St Brieuc bay. In: J. Law (ed.), *Power, Action and Belief. A New Sociology of Knowledge?* (pp. 196–233), Routledge and Kegan Paul: London.

Callon, M. (1991) Techno-economic networks and irreversibility. In: J. Law (ed.), *A Sociology of Monsters? Essays on Power, Technology and Domination, Sociological Review Monograph* (pp. 132–161), Routledge: London.

Callon, M. (1999) Actor-network theory: The market test. In: J. Law and J. Hassard (eds), *Actor Network and After* (pp. 181–195), Blackwell and the Sociological Review: Oxford and Keele.

Callon, M. and Law, J. (1995) Agency and the hybrid collectif. *The South Atlantic Quarterly*, 94 (2), 481–507.

Callon, M. and Law, J. (1997) After the individual in society: Lessons on collectivity from science, technology and society. *Canadian Journal of Sociology*, 22 (2), 165–182.

Callon, M. and Caliskan, K. (2005) New and old directions in the anthropology of markets. Paper presented at the conference New Directions in the Anthropology of Markets, New York University, April 5 2005.

Coakes, E. (2003) Socio-technical thinking: An holistic viewpoint. In: S. Clarke, E. Coakes, M.G. Hunter, and A. Wenn (eds), *Socio-Technical and Human Cognition Elements of Information Systems* (pp. 1–4), Information Science Publishing: Hershey.

Conchie, S.M. and Donald, I.J. (2008) The functions and development of safety specific trust and distrust. *Safety Science*, 46 (1), 92–103.

Corritore, C. L., Kracher, B. and Wiedenbeck, S. (2003) On-line trust: Concepts, evolving themes, a model. *Trust and technology [Special issue]. International Journal of Human-Computer Studies*, 58 (6), 737–758.

Cox, S., Jones, B. and Collinson, D. (2006) Trust relations in high reliability organizations. *Risk analysis*, 26 (5), 1123–1138.

Crawford, T.H. (1993) An interview with Bruno Latour. *Configurations*, 1 (2), 247–268.

Czarniawska, B. and Hernes, T. (eds) (2005) *Actor-Network Theory and Organizing*. Elanders Berlings: Malmo.

Daly, J., Kellehear, A. and Gliksman, M. (1997) *The Public Health Researcher: A Methodological Approach*. Oxford University Press: Melbourne, Australia.

De Vries, P.W. (2005) 'Trust in Systems: Effects of Direct and Indirect Information', Proefschrift, Technische Universiteit Eindhoven, Eindhoven.

Dolwick, J.S. (2009) 'The social' and beyond: Introducing actor-network theory. *Journal of Maritime Archaeology*, 4 (1), 21–49.

Dzindolet, M. T., Peterson, S. A., Pomranky, R. A., Pierce, L. G. and Beck, H. P. (2003) The role of trust in automation reliance. *International Journal of Human Computer Studies*, 58 (6), 697–718.

Edson, R. (2008) *Systems Thinking. Applied: A Primer*. ASysT Institute. http://www.asysti.org/OurModules/SharedDocs/HistoryDocView.aspx?ID=373 (accessed 1 February 2010)

Famewo, J., Matthews, M. and Lamoureux, T. (2007) Model of information aggregation pertaining to combat identification: A review of the literature. *DRDC Toronto No. CR-2007–062*.

Faraj, S., Kwon,D. and Watts, S. (2004) Contested artifact: Technology sensemaking, actor networks, and the shaping of the web browser. *Information Technology and People*, 17 (2), 186–209.

Garson, G.D. (2009) Computerized simulation in the social sciences: A survey and evaluation. *Simulation and Gaming*. 40 (2), 267–279.

Gartner, J. and Wagner, I. (1996) Mapping actors and agendas: Political frameworks of systems design and participation. *Human-Computer Interaction*, 11 (3), 187–214.

Gherardi, S. and Nicolini, D. (2000) The organizational learning of safety in communities of practice. *Journal of Management Inquiry*, 9 (1), 7–18.

Giffin, K. (1967) The contribution of studies of source credibility to a theory of interpersonal trust in the communication process. *Psychological Bulletin*, 68 (2), 104–20.

Graham, S. (1998) The end of geography or the explosion of place? Conceptualizing space, place and information technology. *Progress in Human Geography*, 22 (2), 165–185.

Harbers, H. (2005) *Inside the Politics of Technology: Agency and Normativity in the Co-production of technology and Society*. Amsterdam University Press: Amsterdam.

Hawley, J.K. and Mares, A.L. (2007) *Developing Effective Adaptive Missile Crews and Command and Control Teams for Air and Missile Defense Systems*. Army Research Laboratory (ARL-SR-149).

Hawley, J.K. Mares, A.L. and Giamannco, C.A. (2005) *The Human Side of Automation: Lessons for Air Defense Command and Control.* Army Research Laboratory Technical Report (ARL-TR-3468).

Hawley, J.K., Mares, A.L. and Marcon, J.L. (2010) On fratricide and the operational reliability of target identification decision aids in combat identification. In: D.H Andrews, R.P. Herz and M.B. Wolf (eds), *Human Factors Issues in Combat Identification* (pp. 299–312), Ashgate Publishing: Farnham, Surrey.

Hernes, T. (2005) The organization as a nexus of institutional macro actors: The story of a lopsided recruitment case. In: B. Czarniawska, and T. Hernes (eds), *Actor-Network Theory and Organizing* (pp. 112–128), Elanders Berlings: Malmo.

Jamieson, G.A., Wang, L. and Neyedli, H.F. (2008) *Developing Human-Machine Interfaces to Support Appropriate Trust and Reliance on Automated Combat Identification Systems*. Defence Research and Defence Canada Toronto (CR-2008–114).

Jeffcott, S., Pidgeon, N., Weyman, A. and Walls, J. (2006) Risk, trust, and safety culture in UK train operating companies. *Risk Analysis*, 26 (5), 1105–1121.

Jones, G. R. and George, J. M. (1998) The experience and evolution of trust: Implications for cooperation and teamwork. *Academy of Management Review*, 23 (3), 531–546.

Kramer, R. M. and Tyler, T. R. (eds) (1996) *Trust in Organizations: Frontiers of Theory and Research.* Sage Publications: Thousand Oaks, CA.

Ladkin, P.B. and Stuphorn, J. (2004) Two causal analyses of the Black Hawk shootdown during Operation Provide Comfort. In: P. Lindsay and T. Cant (eds), *Proceedings of the Eighth Australian Workshop on Safety Critical Systems and Software (SCS 2003)* (pp. 3–23), CRPITACS: Cranberra, Australia.

Lanzara, G.F. and Morner, M. (2005) Artifacts rule! How organizing happens in open source software projects. In: B. Czarniawska, and T. Hernes (eds), *Actor-Network Theory and Organizing* (pp. 67–90), Elanders Berlings: Malmo.

Latour, B. (1987) *Science in Action: How to Follow Scientists and Engineers Through Society*. Open University Press: Milton Keynes.

Latour, B. (1988) *The Pasteurization of France.* Harvard University Press: Cambridge, MA.

Latour, B. (1992) Where are the missing masses? The sociology of a few mundane artifacts. In: W.E. Bijker and J. Law (eds), *Shaping Technology – Building Society: Studies in Sociotechnical Change* (pp. 225–258), MIT Press: Cambridge, MA..

Latour, B. (1993) *We Have Never Been Modern.* Harvester Wheatsheaf: Hemel

Latour, B. (1994a) On technical mediation: Philosophy, sociology, geneology. *Common Knowledge*, 3, 29–64.

Latour, B. (1994b) Pragmatogonies: A mythical account of how humans and nonhumans swap properties. *Behavioural Scientist*, 37 (6), 791–808.

Latour, B. (1996) On actor-network theory. A few clarifications. *Soziale Welt*, 47, 369–381.

Latour, B. (1999a) *Pandora's Hope: Essays on the Reality of Science Studies*. Harvard University Press: Cambridge, MA.

Latour,B. (2005) *Reassembling the Social: An Introduction to Actor Network Theory*. Oxford University Press: Oxford.

Law, J. (1987) Technology and heterogeneous engineering: The case of Portuguese expansion. In: W.E. Bijker, T.P. Hughes, and T.J. Pinch (eds), *The Social Construction of Technological Systems: New Directions in the Sociology and History of Technology* (pp. 111–134), MIT Press: Cambridge, MA.

Law, J. (1992) Notes on the theory of the actor network: Ordering, strategy and heterogeneity. *Systems Practice*, 5 (4), 379–393.

Law, J. (1994) *Organizing Modernity*. Blackwell Publishers: Oxford.

Law, J. (1999) After ANT: complexity, naming and topolog': In: J. Law and J. Hassard (eds), *Actor Network Theory and After*(pp. 1–14), Blackwell: Oxford.

Law, J. (2000) *Ladbroke Grove, or How to Think about Failing Systems.* Centre for Science Studies, Lancaster University, Lancaster LA1 4YN, UK, http://www.comp.lancs.ac.uk/sociology/papers/Law-Ladbroke-Grove-Failing- Systems.pdf ,(accessed 1 May 2005).

Law, J. and Callon, M. (1988) Engineering and sociology in a military aircraft project: A network analysis of technological change. *Social Problems*, 35 (3), 284- 297.

Lee, J.D. and See, K.A. (2004) Trust in automation: Designing for appropriate reliance. *Human Factors*, 46 (1), 50–80.

Leveson, N. G. (2002) *System Safety Engineering: Back to the Future*, Aeronautics and Astronautics Department, Massachusetts Institute of Technology, Cambridge, MA

Leveson, N., Allen, P. and Story, M. (2002) The analysis of a friendly fire accident using a systems model of accidents. In: *Proceedings of the 20th International System Safety Conference, Denver, Colorado, 5–9 August 2002*.

Lippert, S.K. and Swiercz, P.M. (2005) Human resource information systems (HRIS) and technology trust. *Journal of Information Science*, 31 (5), 340–353.

Masys, A.J. (2010) 'Fratricide in Air operations: Opening the Black-box – Revealing the Social'. PhD dissertation, University of Leicester, UK.

Masys, A.J., Roza, M., Giannoulis, C., Jacquart, R. (2008) Verification, Validation and Accreditation (VV&A): The GM-VV contribution and roadmap. In: *Proceedings of NMSG 60 Symposium: 'How is Modeling and Simulation Meeting the Defence Challenges out to 2015?', Vancouver, BC, 7–8 October 2008*.

Michael, M. (2003) Between the mundane and the exotic: Time for a different sociotechnical stuff. *Time and Society*, 12 (1), 127–143.

Milewski, A. E. and Lewis, S. H. (1997) Delegating to software agents. *International Journal of Human Computer Studies*, 46 (4), 485–500.

Ministry of Defence (MOD) (2004) *Military Aircraft Accident Summary: Aircraft Accident to the Royal Air Force Tornado GR MK4A ZG710*. Directorate of Air Staff, Metropole Building, Northumberland Avenue, London WC2N 5BP.

Monteiro, E. (2000) Actor network theory and information infrastructure. In: C. Ciborra (ed.), *From Control to Drift. The Dynamics of Corporate Information Infrastructures* (pp. 71–83), Oxford University Press: Oxford. www.idi.ntnu.no/~ericm/ant.FINAL.htm, (accessed 11 July 2005).

Muir, B.M. (1987) Trust between humans and machines, and the design of decision aids. *International Journal of Man-Machine Studies*, 27 (5–6), 527–539.

Muir, B.M. (1994) Trust in automation. Part I. Theoretical issues in the study of trust and human intervention in automated systems. *Ergonomics*, 37 (11), 1905–1922.

Muir, B.M. and Moray, N. (1996) Trust in automation. Part II. Experimental studies of trust and human intervention in a process control simulation. *Ergonomics*, 39 (3), 429–460.

National Audit Office (NAO) (2006) *Ministry of Defence: Progress in Combat Identification.* Report by the Comptroller and Auditor General, HC 936 Session 2005–2006: 3 March 2006.

Parasuraman, R. and Riley, V. (1997) Humans and automation: Use, misuse, disuse, abuse. *Human Factor*, 39 (2), 230–253.

Patriot System Performance: Report Summary (2005) *Report of the Defence Science Board*, http://www.acq.osd.mil/dsb/reports/2005–01-Patriot_Report_Summary.pdf.

Rempel, J. K., Holmes, J. G. and Zanna, M. P. (1985) Trust in close relationships. *Journal of Personality and Social Psychology*, 49 (1), 95–112.

Riegelsberger, J. (2005) 'Trust in Mediated Interactions'. PhD dissertation, University College London, London.

Riegelsberger, J., Sasse, M.A. and McCarthy, J.D. (2005) The mechanics of trust: A framework for research and design. *International Journal of Human-Computer Studies*, 62 (3), 381–422.

Rolland, K.H. and Aanestad, M. (2003) The techno-political dynamics of information infrastructure development: Interpreting two cases of puzzling evidence. Presented at IRIS-26 Porvoo, Finland, August 9–12 2003. http://heim.ifi.uio.no/~margunn/2003/IRIS26.pdf (accessed 23 May 2007).

Roman, P.A. (2005) Garbage in, Hollywood out! *Proceedings SimtecT, 2–6 May 2005, Sydney Australia.*

Rotter, J. B. (1967) A new scale for the measurement of interpersonal trust. *Journal of Personality*, 35, 651–665.

Schobel, M. (2009) Trust in high-reliability organizations. *Social Science Information*, 48 (2), 315–333.

Senge, P. (1990) *The Fifth Discipline: The Art and Practice of the Learning Organization.* Doubleday Currency: New York.

Snook, S.A. (2000) *Friendly Fire: The Accidental Shootdown of US Black Hawks over Northern Iraq.* Princeton University Press: Princeton, NJ.

Somerville, I. (1997) Actor network theory: A useful paradigm for the analysis of the UK cable/on-line socio-technical ensemble? Available at: http://hsb/baylor.edu/eamsower/ais.ac.97/papers/somervil.html [accessed 10 August 2004].

Somerville, I., (1999) Agency versus identity: Actor-network theory meets public relations. *Corporate Communications: An International Journal*, 4 (1), 6–13.

Stanton, N. (2010) Situation awareness: Where have we been, where are we now and where are we going? *Theoretical Issues in Ergonomics Science*, 11 (1–2), 1–6.

Stanton, N.A., Salmon, P.M., Walker, G.H. and Jenkins, D.P. (2009) Genotype and phenotype schemata and their role in distributed situation awareness in collaborative systems. *Theoretical Issues in Ergonomics Science*, 10 (1), 43–68.

Tatnall, A. 2000. 'Innovation and Change in the information systems curriculum of an Australian university: A socio-technical perspective'. PhD dissertation, Central Queensland University.

Thrift, N. (1996) New urban eras and old technological fears: Reconfiguring the goodwill of electronic things. *Urban Studies*, 33 (8), 1463–1494.

Van der Duim, V.R. (2005) 'Tourismscapes: An Actor-Network Perspective on Sustainable Tourism Development'. Dissertation, Wageningen University, Wageningen.

Verbeek, P. (2005) Artifacts and attachment: A post-script philosophy of mediation. In: H. Harbers, (ed.), *Inside the Politics of Technology: Agency and Normativity in the Co-production of Technology and Society* (pp. 125–146), Amsterdam University Press: Amsterdam.

Viseu, A.A.B. (2005) 'Augmented Bodies: The Visions and Realities of Wearable Computers'. PhD dissertation, University of Toronto, Toronto.

Wang, L., Jamieson, G.A. and Hollands, J.G. (2008) Improving reliability awareness to support appropriate trust and reliance on individual combat identification systems. In: *Proceedings of the Human Factors and Ergonomics Society 52nd Annual Meeting 2008* (pp. 292–296).

Weick, K.E. and Sutcliffe, K.M. (2007) *Managing the Unexpected: Resilient Performance in an Age of Uncertainty* (2nd edn). John Wiley and Sons Inc.: San Francisco, CA.

Wickramasinghe, N., Tumu, S., Bali, R.K. and Tatnall, A. (2007) Using Actor Network Theory (ANT) as an analytic tool in order to effect superior PACS implementation. *International Journal of Networking and Virtual Organizations*, 4 (3), 257–279.

William-Jones, B. and Graham, J.E. (2003) Actor-Network Theory: a tool to support ethical analysis of commercial genetic testing. *New Genetics and Society*, 22 (3), 271–296.

Woods, D.D. and Cook, R.I. (1999) Perspectives on human error: Hindsight biases and local rationality www.nifc.gov/safety/.../Perspectives%20on%20Human%20Error.pdf (accessed 15 December 2009).

Woods, D., Johannesen, L., Cook, R. and Sarter, N. (1994) *Behind Human Error: Cognitive Systems, Computers, and Hindsight*. Crew Systems Ergonomics Information Analysis Center (CSERIAC), Wright-Patterson Air Force Base, OH, December, 1994.

Yeung, H.W.C. (2002) Towards a relational economic geography: Old wine in new bottles? Paper presented at the 98th Annual meeting of the Association of American Geographers, Los Angeles, CA. 19–23 March 2003. http://courses.nus.edu.sg/course/geoywc/publication/Yeung_AAG.pdf

# Author Index

# Subject Index

For Product Safety Concerns and Information please contact our EU representative GPSR@taylorandfrancis.com
Taylor & Francis Verlag GmbH, Kaufingerstraße 24, 80331 München, Germany

www.ingramcontent.com/pod-product-compliance
Ingram Content Group UK Ltd.
Pitfield, Milton Keynes, MK11 3LW, UK
UKHW020015200926
475702UK00014B/1897
*9781138075948*